ALSO BY RICHARD MORRIS

The Evolutionists

Achilles in the Quantum Universe

Cosmic Questions

Time's Arrows

The Big
QUESTIONS

The Big
QUESTIONS

Probing the Promise and Limits of Science

Richard Morris

TIMES BOOKS

Henry Holt and Company • New York

Times Books
Henry Holt and Company, LLC
Publishers since 1866
115 West 18th Street
New York, New York 10011

Henry Holt® is a registered trademark of
Henry Holt and Company, LLC.

Library of Congress Cataloging-in-Publication Data

Morris, Richard, date.
    The big questions : probing the promise and limits of science /
    Richard Morris.—1st ed.
        p. cm.
    Includes bibliographical references and index.
    ISBN 0-8050-7092-3 (hb)
    1. Science—Philosophy. I. Title.

    Q175 .M8688 2002
    501—dc21   2001057362

First Edition 2002

Designed by Paula Russell Szafranski

Printed in the United States of America

10 9 8 7 6 5 4 3 2 1

Contents

Preface

According to the British philosopher Alfred North Whitehead, "The safest general characterization of the European philosophical tradition is that it consists of a series of footnotes to Plato." Although this statement should probably be taken with a grain of salt, Whitehead did have a point. It was Plato who invented many of the philosophical questions that philosophers asked over the next two and a half thousand years. Although few people today believe many of Plato's philosophical doctrines, he is still considered to be one of the greatest, if not *the* greatest, philosophers of all time. Some of the questions he asked are still not so easy to answer.

Philosophical questions have always fascinated me, and since I also have an interest in cultural history, I spend a certain amount of time reading the philosophers of past ages. It is always interesting to look at the ways in which the philosophers of those times thought, and to consider the relationships between their ideas and the social and cultural trends of their ages.

One of the philosophers I read with great interest is the American philosopher and psychologist William James, who died in 1910. I don't find myself in agreement with everything he says, but I find that he too asked interesting questions. And sometimes he took a fresh look at some of the questions that had been asked by his predecessors.

One day, when I was reading James's posthumously published *Some Problems of Philosophy*, I came to a chapter where he cited a number of metaphysical questions that philosophy had asked. James considered the questions to be purely philosophical in nature. It never occurred to him that science might be able to answer such queries as "How comes there to be a world at all? And, Might it as well not have been?" Or, "How are mind and body joined? Do they act on each other?" In his opinion, such questions were too fundamental to be part of the subject matter of science.

But as I read through James's list of metaphysical questions, I began to realize that science *was* answering, or attempting to answer, many of the questions he cited. The cognitive sciences were examining some of the old metaphysical questions about the mind. Cosmologists were speculating about how the universe came into being, and physicists working in superstring theory were asking questions about the nature of space and time. Quantum mechanics obviously had something to say about determinism, and also about the nature of "objective reality." Questions that were once considered to be metaphysical in nature were, in our time, becoming objects of scientific study.

It was obvious that science had not yet answered all of these questions. However, a little thought was sufficient to convince me that in most cases it had at least made some progress. Then, slowly at first, an outline for this book began to form in my mind. I realized that quite a bit could be said about scientific approaches to these questions.

Scientists have generally not been drawn to these questions because they were preoccupied with philosophical ideas. Rather, they considered them because they arose in scientific contexts. For example, Einstein introduced new ideas about the nature of space and time, not because he pondered the things that philosophers had said, but because they arose naturally in the context of his theories. Similarly, indeterminism cropped up in quantum mechanics because it was an inevitable consequence of the mathematical assumptions that were made.

Nevertheless the fact that scientists today are wrestling with some of the questions that philosophers have asked over the ages is not an accident. Scientists endeavor to understand the nature of physical reality, and questions about the nature of reality have traditionally been part of philosophy. As a result, science frequently finds itself attempting to answer some very old questions in new ways.

The Big
QUESTIONS

Introduction:
Science and Philosophy

Greek philosophy begins with Thales, who said that everything was made of water. To many people this may not sound like a very impressive beginning. However, the idea isn't as crazy as it sounds. Thales must have noted that evaporation turned water into mist, and that freezing caused it to solidify. Aristotle suggests that he might have gotten the notion "perhaps from seeing that the nutriment of all things is moist, and that heat itself is generated from the moist and kept alive by it. . . . He got his notion from this fact and from the fact that the seeds of all things have a moist nature, and water is the origin of the nature of moist things." To someone pondering what things were made of in a time when there was not yet any such thing as science, it must have seemed a plausible idea.

Thales's writings are lost, and little is known about his life. It is necessary to depend upon later commentators, such as Aristotle, for knowledge of his philosophy. However, Thales can be roughly dated by the fact that he is supposed to have predicted an eclipse of the sun in 585 B.C. He was apparently familiar with the knowledge of the Babylonian astronomers, who knew that eclipses recurred in nineteen-year cycles. The Babylonians could predict lunar eclipses pretty accurately. Predicting solar eclipses is more difficult, since an eclipse is not visible from all areas of the earth. But they did know at what times a solar eclipse *might* be seen.

Thales's idea that everything was made of water was a scientific hypothesis. In some respects, it is reminiscent of the theory held by modern cosmologists that the universe was originally composed of hydrogen and helium. Like modern scientists, Thales wondered about the basic constituents of the world around him. Naturally not everyone found his ideas to be plausible. Thales's ideas about the fundamental constituents of matter were questioned by later philosophers who doubted that his theory could explain the origin of things that were dry. None of them were able to prove that Thales had been wrong. Unlike scientific theories, which depend on empirical evidence, philosophical ideas are rarely conclusively refuted. However, philosophers did attempt to improve on Thales's idea, and eventually the theory proposed by Empedocles that earth, air, fire, and water were the four basic elements came to be accepted. Although the latter theory doesn't seem to be that superior to modern eyes, it appeared quite logical to the ancients.

Thales was not only the first Western philosopher, he was also the first scientist. There is nothing very surprising about this. At the time, science was not distinguishable from philosophy. They did not become separate disciplines until more than two millennia later. Consequently many of Thales's philosophical successors continued to speculate about such things as constituents of matter, terrestrial motion, the nature of heavenly bodies, the nature of time, and other topics that are now thought to fall into the domain of science. To be sure, not every ancient philosopher showed an interest in what we think of as scientific topics. Socrates, for example, was considerably more interested in ethics than he was in the nature of the world. And Plato sometimes exhibited an anti-scientific bias. For example, he believed that students in his Academy should not concern themselves too much with the behavior of actual celestial objects; they should study the mathematics of the motion of ideal heavenly bodies instead. But there were other philosophers who continued the new scientific tradition. The best examples of this are the writings of Aristotle, which deal at length with such topics as meteorology, physics, and biology.

Ancient Greek science was not anything like modern science. For one thing, the Greeks performed no experiments. Nor did they generally depend upon observation. They tended to believe that the nature of the world could be understood by means of thought alone. This is why Plato thought knowledge of actual heavenly bodies to be unimportant. Aristotle paid more attention to the natural world than Plato did, but he developed his physics from general principles, obtaining results that were sometimes ludicrously wrong. According to Aristotle, a projectile does not follow a curved path; on the contrary, it proceeds in a straight line for a certain distance and then drops straight down. Today, every high school physics student knows that it does nothing of the sort. Similarly, Aristotle was wrong when he maintained that heavy bodies fell faster than light ones. But so great was his influence that this continued to be believed until Galileo performed experiments that proved Aristotle to be wrong.

The Greeks cannot be criticized too harshly for their failure to depend upon empirical data. They invented the idea of scientific inquiry, and we can hardly blame them for knowing nothing of modern methods. They were the ones who first thought to ask what everything was made of, and why objects behaved the way that they did. They did this because they asked questions about everything: about the nature of ethics, about the mind, about politics, about aesthetics, and about physics and cosmology.

So influential were their methods of questioning that, more than two thousand years after Thales died, there were still no "scientists," only "natural philosophers." For example, when Newton published his monumental work on gravitation and other principles of physics in 1687, he gave it the Latin title *Philosophiae naturalis principia mathematica* (Mathematical Principles of Natural Philosophy). And no one thought it odd that Newton's contemporaries, the philosophers René Descartes and Gottfried von Leibniz, should be making contributions to both science and philosophy. That was the sort of thing that philosophers did. The word *scientist* did not come into general use until much later, around the middle of the nineteenth century. The date of the first citation for the word in the *Oxford English Dictionary* is 1840.

The Uses (or Lack Thereof) of Philosophy

Rarely are any definitive answers found to philosophical questions when they remain purely philosophical in nature. If it becomes possible to make attempts to answer them empirically, they become scientific questions instead. For example, there was no way that Thales's theory about the nature of matter could have been tested. The relevant evidence simply didn't exist in his day. It was only after a long series of discoveries were made by modern chemists and physicists that it was possible to say what matter was made of. Similarly, it only became possible to answer questions about the nature of celestial bodies when the requisite scientific technology became available.

Noting that philosophy has found few answers to the questions that it has asked, a skeptic might be inclined to concude that it is a pretty useless activity, that it should be regarded as nothing more than a prelude to science. If he did believe this, I think that he would be mistaken. The essence of philosophy is questioning. Thales is important because he was the first to ask what the world was made of. His successors, including Plato and Aristotle, were the first to ask significant questions about the nature and the origin of the cosmos, the nature of mind, the nature of human knowledge, and many other fundamental things. Their ideas were so fertile that scientists and philosophers alike are still attempting to answer some of their questions today.

Science grew directly out of philosophy. Modern experimental science began with Galileo. He was the first to perform experiments and to attempt to describe mathematically the behavior of moving bodies. One of the things that motivated him to do this was his dissatisfaction with the teachings of Aristotle on such matters. Of course, Galileo turned out to be right, and Aristotle was proved to be wrong. Heavy objects do not fall more rapidly than light ones. Projectiles did not move in the way that Aristotle said they did. However, if Aristotle or other philosophers had never contemplated such matters, it might never have occurred to Galileo to think about them. If the relevant questions had never been posed, no one would have been likely to go looking for answers, and modern science might have begun much later than it actually did.

One of the last disciplines to split off from philosophy was psychology. Experimental psychology began around the middle of the nineteenth century. Since modern philosophers had been speculating about psychological matters for nearly two centuries, the first psychologists had the advantage of being heirs to a rich intellectual tradition. Large parts of their research program were already laid out for them. Today most physical scientists, such as physicists, do not consider philosophy relevant to the kinds of research they do. But matters are different in the psychological disciplines, which are still being enriched. As we shall see in a subsequent chapter, philosophical questioning is still enriching the field of cognitive science today. Scientists are beginning to be able to answer questions about the nature of mind partly because philosophy has provided so many important questions and concepts.

Philosophical questioning and scientific questioning are not the same thing. The latter depends upon observation and experiment and the former usually does not. Nevertheless, the asking of questions is fundamental to both disciplines. And often the questions that scientists find most relevant are the ones that were originally proposed by philosophers. We will see that nowadays numerous questions that most people would consider somewhat metaphysical are being tackled by hardheaded scientists.

Big Questions

In *Some Problems of Philosophy*, the American philosopher and psychologist William James devoted a chapter to "The Problems of Metaphysics." Rather than try to define the word *metaphysics*, James gave some examples of metaphysical questions. Included in his list were the following:

> What are "thoughts" and what are "things" and how are they connected?
> What do we mean when we say "truth"?

How comes there to be a world at all? And, Might it as well not
  have been?
What is the most real kind of reality?
What binds all things into one universe?
Is unity or diversity more fundamental?
Have all things one origin? Or many?
Is the world infinite or finite in amount?
How are mind and body joined? Do they act on each other?
How does anything act on anything else?
How can one thing change or grow out of another thing?
Are space and time beings?—or what?
In knowledge, how does the object get into the mind?—or the
  mind get at the object?

"Such are specimens of the kind of question termed metaphysical,"
James says. Now, most of these are not the kinds of questions that
philosophers ask today; constructing metaphysical systems is no longer
the fashion. However, they are similar to the kinds of questions that
contemporary scientists are attempting to answer. Philosophical questions are becoming scientific questions today, just as they were in the
day of Galileo, or of the early experimental psychologists.

Cognitive science, which is an amalgam of such fields as cognitive
psychology, neurophysiology, and artificial intelligence, is beginning to
answer some of the old philosophical questions about the nature of the
mind. Meanwhile cosmologists and theoretical physicists often speculate about the reasons why our universe came into being. The big bang
theory of the origin of the universe has now been accepted for decades.
But that theory does not tell us why there was a big bang to begin with.
The development of the new fields of quantum cosmology and superstring cosmology, however, has allowed scientists to pose fundamental
questions about the origin of our universe. They are even beginning to
ask if there might not be other universes, perhaps an infinite number
of them.

Questions regarding the nature of time have long been among the
outstanding unsolved problems of physics. It is now generally agreed
that space and time were created in the big bang. But time has always

been a puzzle. Unlike space, it has a preferred direction. We travel from the present toward the future, not the past. The reason this is puzzling is that the fundamental laws of physics do not have a preferred time direction; they would work equally well if time went "backwards." "What is time?" was once a philosophical query. Nowadays it has become an important scientific question, one that has not yet been completely answered.

Since the time of Socrates, philosophers have been asking questions about the human condition, what it means to be "human." Nowadays, many of these very same questions are being asked not by philosophers, but by evolutionary psychologists and other scientists. Our "human nature," after all, is something that evolved at the same time evolution was shaping our bodies. Scientists are beginning to speculate about the evolutionary reasons for certain typical human behavior patterns.

Many of the metaphysical questions that philosophers once asked are becoming a part of scientific inquiry. It has been discovered it is possible to examine some of them empirically. In other words, they are no longer philosophical problems, they are science. As the frontiers of scientific knowedge are pushed outward, new realms of inquiry are being discovered and we are beginning to understand things once thought to be "beyond science."

For example, as recently as 1960 most physicists believed that questions about the origin of the universe would never be answered. Consequently, they did not take hypotheses like the big bang theory very seriously. Today they are confident they understand what must have been happening when the universe was only a tiny fraction of a second old. Thus, they have begun to speculate, not only about the creation of the universe but also about such matters as the origin of space and time and the reasons why there are three dimensions of space. They are posing questions that no self-respecting scientist would have thought of asking forty or fifty years ago.

Similarly, scientists are now asking questions about the nature of the mind that few scientists would have dared pose forty or fifty years ago. Behaviorist psychology, which emphasizes stimulus and response, and which tried to do away with such "mentalist" concepts as "thinking," was still dominant in American psychology during the 1950s and 1960s.

But the revolution in cognitive science that took place during the second half of the twentieth century has changed matters dramatically. Cognitive scientists are also asking questions that a previous generation of scientists might not even have understood.

There have been equally astonishing advances in such diverse disciplines as molecular biology, quantum mechanics, and the physics of elementary particles, neurophysiology, and evolutionary psychology. In these fields also, what were once metaphysical questions have become objects of scientific inquiry. As the new millennium begins, there is still much that is not yet understood. But one of the reasons for this is that the horizons of science have broadened tremendously. As a result we are likely to see some astonishing discoveries in the years ahead.

In the chapters that follow, I will pose some of James's "metaphysical" questions and examine them from the standpoint of current scientific knowledge. I will try to see to what extent science has answered them, or is trying to answer them. Since these are the "big questions," it is to be expected that in many cases scientific understanding of them is incomplete. However, I think I will be able to show that, in almost every case, science is able to at least shed some light on these questions, even the most baffling of them.

For example, one of the questions James asked, "Why is there a world at all?," was first posed by the German philosopher Gottfried von Leibniz approximately three centuries ago. Of all philosophical questions, this would seem to be the most difficult to answer, or so it seemed until the concluding decades of the twentieth century. Then, suddenly, everything changed. Physicists discovered that quantum mechanics might be capable of answering this question in a fairly straightforward way. To be sure, the theories that these physicists proposed were speculative. But the question was placed on a firm scientific foundation, and it was shown that it was possible to find plausible answers to what had seemed an impossibly difficult question.

Philosophers have speculated about the nature of the material world for millennia. There have always been some who denied its objective reality in one sense or another. For example, Plato taught that the world of the senses was less real than an ideal world which could be apprehended only by thought. The European idealistic philosophers denied the material world altogether, maintaining that nothing existed but mind.

Such ideas sound antithetical to the scientific outlook. By definition, physical science deals with real material objects. After all, it would not be possible to study the material world if it did not really exist. However, there is one scientific field in which we find scientists denying the objective reality of the objects they study. If one accepts the most commonly believed interpretation of quantum mechanics, subatomic particles such as protons, neutrons, and electrons do not have such objectively real properties as position, momentum, or energy until those properties are observed. Until they are observed, they are only fuzzy bundles of probabilities. There does seem to be a sense in which the British philosopher George Berkeley was right when he maintained that nothing was real until it was perceived.

I will be discussing many questions in the chapters that follow. You may find some of the answers that scientists have suggested to be surprising. Others will seem to be speculative. However, you should remember that this is informed speculation. Scientists are subject to strong constraints when they speculate. They cannot ignore well-established empirical facts, and their ideas must be made to seem scientifically plausible. For example, physicist Alan Guth has suggested that a sufficiently advanced technological civilization might be able to create new universes at will. But this is not an off-the-cuff comment. Guth has studied the problem of the manner in which this might be done in great detail. When he says that it is conceivable that new universes might be created, he gives a detailed explanation of how this might be done.

And, of course, by doing this he suggests a possible answer to one of the big questions.

1 | *What Is Time?*

Near the end of the fourth century, St. Augustine began to ponder the nature of time. The more he thought about it, the more puzzled he became. "What is time?" he wrote. "If no one asks me, I know what it is. If I wish to explain to him who asks, I do not know." Augustine looked to see what Aristotle had said. But Aristotle's writings on time, he found, were of little help. Aristotle had called time the "number of motion." Augustine realized that if one were to measure motion in terms of time, and time in terms of motion, nothing was explained. "Do I measure, O my God, and know not what I measure?" Augustine asked despairingly.

Of course, the story has a happy ending. Whenever a philosopher professes to be completely baffled by an idea, we can generally be sure that some ingenious theory about the matter will not be long in coming. Augustine was no exception. After expressing all these doubts, he presented a somewhat subjective theory of time. Time, he said, was something that resided in the soul. From this it seems to follow that, if there were no human beings, there would be no time. And indeed this is what

Augustine concluded. He held that God resided in Eternity, which was outside of time, and that there had been no time before God created Earth and the first human beings.

Augustine's subjective view of time has a certain plausibility. We are all aware of the passage of time, and of the fact that we remember the past and await the future. However, one can't help but think that there must be more to time than this. If human beings did not exist, there would presumably still be time. A day would still have the same length, and Earth would still revolve around the sun in a year. Furthermore, time must have existed during the billions of years that passed before human beings evolved. The universe, which is around 13 or 14 billion years old, has a long history, and we have been here only a short while.

But it is not so easy to say precisely what time is. We measure time by the movement of Earth around the sun. But, as Augustine pointed out, how can one measure motion except in terms of time? And what exactly is the passage of time? Is time something that "flows" past us? Or is it more like a river that carries us from the present toward the future? Did time have a beginning? Can it come to an end? And what *is* the flow of time? It certainly isn't anything that can be measured in the laboratory. Scientists can use accurate atomic clocks to measure the time interval between two events to a high degree of precision. But no instrument can detect time "flowing." It appears that if we want to talk about the flow of time all we can say is that it moves at a rate of one second per second. This is hardly meaningful.

The problem of the nature of time is not an easy one. Contemplating it has caused some philosophers, such as Parmenides, Plato, Spinoza, and Hegel, to conclude that the concept of time contained serious contradictions, and that therefore time couldn't be real. I don't know about you, but I can't help but feel that this is an evasion. Or at least an attempt to solve a problem by making the claim that the thing that is problematic doesn't really exist.

One of the philosophers who denied the reality of time was the Cambridge philosopher J. Ellis McTaggart (the "profound McTaggart" mentioned in William Butler Yeats's poem "A Bronze Head"). In 1908 McTaggart published an article in the philosophical journal *Mind* in which he presented his arguments on the topic. I think these arguments are worth describing in a little detail, not because I think there are any good reasons to believe them, but because they provide a good example of the kinds of problems that arise when one begins to ponder the nature of time.

McTaggart presented two descriptions of time, which he called the A-series and the B-series. By A-series he meant the description of time that depends upon the ideas of past, present, and future. By B-series, he meant the difference between earlier and later. As McTaggart pointed out, these two ways of talking about time are somewhat different from one another. For example, "Elizabeth II is the present queen of England" depends upon the idea of "now," while the statement "The reign of Henry VIII preceded that of Elizabeth I" does not. The first makes reference to the present moment of time. The second is as true today as it was during the reign of Elizabeth I, and it will still be true a thousand years from now.

McTaggart claimed that the two descriptions of time were incompatible. The description of time that depends upon the idea of a moving "now" cannot be reconciled with the idea of events that have fixed dates. Time cannot be both moving and static. Therefore it cannot exist.

I don't know about you, but I'm not convinced by all this. On the other hand, I'm not sure where the flaw in the argument lies. There *is* something mysterious about time. And time is all the more mysterious because subjective time seems to have very little to do with the time of physics.

Time in Physics

At first, time as it is conceived by physicists seems a perfectly simple and straightforward thing. Time is a dimension that is analogous to the three dimensions of space. If we want to specify when and where an event takes place, we need only give its time and space coordinates. For example, suppose I have a dentist appointment. If I want to keep the appointment I must know where it is located in space and in time. For example, the dentist's office may be on the fourth floor of a building at the corner of Green and Virginia Streets. That specifies its location on a two-dimension grid (intersection of two streets) and tells me where it is in the third dimension (fourth floor). Finally, a coordinate in the time dimension (3:00 P.M. on Thursday) tells me when I must be there.

There are three dimensions of space, and one of time. That seems simple enough. To be sure, the term "spacetime" is often heard in connection with Einstein's theories of relativity. But there is nothing very complicated about that. In relativity space is still three-dimensional and time one-dimensional, just as they are in Newton's mechanics. However, space and time often interact with each other, so it is more convenient to consider all four dimensions together. Separating them would make matters more complicated.

At first glance, the time of physics doesn't seem mysterious at all. But as soon as we consider the fact that time has a direction, matters become more problematic. There are no preferred directions in space. On the other hand, time involves the concepts of past and future. Past events can affect events in the present, and present events have influences that will be seen in the future. The reason that this is a problem is that the fundamental laws of physics, including those of Newtonian mechanics, electricity, and magnetism, relativity and quantum mechanics, do not distinguish between past and future. They are *time symmetric*. In other words, they would be equally valid if the direction of time were reversed.

If a videotape were made of the motion of the planets in some distant star system, for example, you would not be able to tell whether the tape was being played forward or backwards, no matter how much physics you knew. If the planets were moving backwards in time, their motion would still be accurately described by the same laws of gravity. They would simply appear to be moving in the opposite direction. Similarly, you would not be able to tell whether a videotape of the collision of two billiard balls or of two or more subatomic particles was being run in the correct direction. The laws that govern such collisions remain the same under time-reversal also.

There are of course some circumstances under which you would be able to tell whether or not the tape was being played correctly. If you saw a tape that showed billiard balls coming together to form a triangle at the center of a pool table, or a type of shattered glass coming together as it rises from the floor to the edge of a table, you would know that something was wrong. However, strange as it sounds, the behavior of the objects in these tapes would not violate any fundamental physical laws. If the billiard balls were moving at just the right velocities in precisely the right directions, they could come together into a motionless triangular group. So what you saw in the backwards-played tape would not be impossible, just highly improbable. Similarly, if the molecules in the broken glass, and in the air around it, had just the right motions, the pieces of the glass could assemble themselves. If the molecules in the floor then cooperated to give it just the right kind of push, it could jump up to the table. To be sure, such a thing is so improbable that even if you waited a time equal to many times the present age of the universe, it is unlikely that you would have much of a chance to see it happen. But again, there is nothing in the basic laws of physics that says that it is impossible.

Time and Probability

It appears, then, that the direction of time might have something to do with probability. But what? Well, fortunately, there is a very precise way of discussing this matter. It is possible to say that the forward direction of time is the direction of increasing entropy. Entropy can be described as the degree of disorder in a system. For example, suppose we put an ice cube in a glass of water and watch it gradually melt. As the melting takes place, the ordered crystalline structure of the ice disappears, and the molecules that made up the ice begin to move randomly in the water. Similarly, when a cue ball strikes the group of billiard balls that have been arranged in a triangle in the middle of a table, the orderly arrangement is broken up. When the balls come to rest, they will be scattered around the table in random positions.

This is not the only way that entropy can be described. It can also be said that entropy is a measure of available energy. Here the key word is "available." For example, the water in a high mountain lake has a certain amount of gravitational energy. But if the lake has no outlet, this energy cannot be used. However, if a river does run out of the lake, the flowing water can be used to turn a water wheel. Alternatively, the river can be dammed, and water running downward through the dam can be used to generate electrical power. All this is possible because the water is able to fall from one level to another.

Similarly, the earth contains a lot of heat energy. Such energy is present in any object that is not at a temperature of absolute zero (the lowest possible temperature, at which all molecular motion ceases; on the Celsius scale it is −273°). But no one has ever devised a way, for example, to extract any heat energy from the earth's oceans. Heat will flow only when there is a difference in temperature between one object and another. Then the energy in the hot object can be used to do work. This is the principle on which gasoline engines work. The hot gases in the cylinder push on the piston and lose heat as the piston is moved. At the end of the cycle, the entropy of the engine has increased, and it can do no more work until more fuel is injected into the cylinder and ignited.

Life on earth depends upon the flow of heat from the sun to the earth. This is also the source of most of the energy we use. Plants convert the energy in sunlight into chemical energy by photosynthesis. Some of this energy is made use of by animals that eat the plants. Thus the energy flow from the sun allows animals to live and reproduce, and is also the source of the energy in our fossil fuels. Sunlight causes water to evaporate from the oceans. It can later fall as rain or snow at high altitudes and provide hydroelectric power as it flows downward toward sea level.

The sun is very hot. The core of the sun, where the nuclear reactions that produce solar energy take place, has a temperature of about 15 million degrees Celsius. Since the earth is relatively cool, a great deal of useful energy can flow from the sun, and the entropy of the sun-earth system is low. However, in about 5 billion years, the nuclear fuel that the sun uses to create energy will begin to be exhausted. Eventually, the sun will evolve into a dim white dwarf star, which will continue to glow only because it retains some residual heat, like the glowing coals that remain after a wood fire that has burned itself out. At that time, there will be little energy flowing from the sun to the earth (if the earth still exists!), and the entropy of the system consisting of the sun and the earth will be high. Little usable energy will be available, and if life on the planet still exists, it will have to find other sources of energy if it is to continue to survive.

As time passes, the entropy of the entire universe increases in a similar manner. Stars grow old and die. Or, if they are massive enough, they collapse into black holes. Star formation is still going on in our universe. But eventually the hydrogen and helium gas from which stars are formed will become exhausted, and no new stars will be created. This process will continue inexorably until the entire universe is a cold, dark place. At that time, the entropy of the universe will be near a maximum.

Entropy can be used to define the direction of time. The universe has evolved from a very low entropy state just after the big bang to the state of higher entropy that exists today. As the stars die and the night sky turns dark, the entropy will become higher yet. At the same time, the universe will continue to expand, and the matter in it will become more dispersed. Eventually life, which depends on the existence of available energy, will become extinct.

Many people find this prospect depressing, even though these events will happen billions of years in the future. This is reason enough to ask whether there might not be any escape from the "heat death" of the universe, whether there could not be some way that life could continue.

The answer is that there might be if only the universe does not continue to expand forever. Remember that the law of increasing entropy depends on probabilities. Theoretically, rare random events can cause the entropy of a system to decrease if one waits long enough. If a sufficiently huge number of billiard games were played, someone would eventually see what were apparently randomly moving balls assemble themselves into a motionless triangle. Now, even if most of us spent most of our time hitting balls with billiard cues, this would not be very likely to happen any time between now and the time that the sun becomes a white dwarf. But it would at least be a theoretical possibility.

Similarly, the possibility of a measurable entropy decrease in the universe at some time or another cannot be ruled out. However, the larger a system is, the more improbable such an event becomes. If the universe continues to exist for an infinite period of time, there will have to be some kind of entropy decrease sooner or later. However, the expansion of the universe ensures that, if this does happen, matter will be so dispersed that this fact is of little significance. Furthermore, most entropy decrease would be tiny blips in the overall pattern of increasing entropy. The law of increasing entropy—called the Second Law of Thermodynamics—is based on probability, and the laws of probability tell us that small fluctuations are much more likely than large ones.

The Five Arrows of Time

The concept of entropy can be used to define an *arrow of time*. As we have seen, the future is the direction in which entropy increases. But this is only one of five ways in which we can make a distinction between past and future. In all there are five arrows of time. These are:

1. Increasing entropy.
2. The expansion of the universe. The galaxies and clusters of galaxies that make up our universe are receding from one another. In the future, the distances between them will be greater than they were in the past.
3. The kaon. In general, reactions between subatomic particles are time reversible. There is one known exception: those involving a particle called the *kaon*.
4. The electromagnetic arrow. Electromagnetic radiation, such as light, ultraviolet radiation, radio waves, and gamma rays, travels toward the future, not toward the past.
5. The psychological arrow of time.

The Expanding Universe

In 1929 the American astronomer Edwin Hubble made one of the most significant discoveries of the twentieth century. He found that most of the galaxies that could be observed were receding from our own galaxy, the Milky Way, at high velocities. Furthermore, the farther away the galaxies were, the greater the speed of their recession.

Hubble's observations naturally did not imply that everything in the universe was flying away from the earth. On the contrary, his findings implied that the galaxies were receding from one another. It was a situation analogous to the rising of a loaf of raisin bread dough. If the raisins represent galaxies, then they will all recede from each other as the dough rises. And the raisins that are farthest from each other to begin with will move away from each other most rapidly.

Like all analogies, this one is somewhat imperfect. A loaf of bread has boundaries. And, according to Einstein's general theory of relativity, the universe doesn't. However far and fast it might go, a spaceship could no more reach the "edge of the universe" than we could travel to the "ends of the earth." However, as long as this is kept in mind, the raisin bread analogy works reasonably well.

When scientists speak of the expanding universe, they mean that the galaxies and clusters of galaxies in the universe are becoming more dispersed. They do not picture the universe—which might be infinite—as an expanding sphere. It is the relative motion of galaxies that is the important thing.

Gravity acts as a braking force on the expansion. It acts to slow down the rate at which galaxies recede from one another. If no other forces are at work (as we shall soon see, this is a big "if"), then it is conceivable that gravity might eventually bring the expansion to a halt. If it does, the universe will begin to contract. Furthermore, it will contract faster and faster as the galaxies become closer together again.

This has caused some scientists to wonder if time might have a different direction in a contracting universe. The idea of backwards time, by the way, is not a new one. Plato speculated about the matter two and a half millennia before anyone knew that there were galaxies. In his dialogue *Statesman*, Plato speculated about a time when the heavens revolved in a direction opposite to the sense in which they revolved now. At that time, according to a story told by one of the characters in the dialogue, time flowed backwards, too, and human beings grew younger rather than older. They grew into children, then into infants, and finally disappeared altogether. There is no reason to think that Plato actually believed that this had ever actually happened. Many of the opinions expressed by Socrates in the dialogues are actually Plato's own ideas. However, this is a story that is told *to* Socrates. Nevertheless, Plato may have considered it to be a possibility. At least Socrates makes no objections to the idea.

In the 1960s the American astrophysicist Thomas Gold produced a scientific version of this theory. Gold reasoned that, in an expanding universe, heat flowed from the stars into space, and the arrow of time pointed toward the future. However, if at some point it began to contract, the stars would have been shining for so long that space would contain a lot of heat and light. The universe would be so hot that, once the contraction began, heat would begin to flow from space into solid objects, heating them in the process. Under such circumstances, the stars would no longer give off heat and light; they would absorb them. This would cause the arrow of time to point in the opposite direction.

If the contracting phase of Gold's universe were inhabited, human beings and animals would grow younger, not older, just as they would in Plato's story. But they wouldn't know that they were growing younger, since all their mental processes would also run in reverse. Their experience of the passage of time would be exactly like ours. What we think of as the far future would be the past to them.

There are difficulties with Gold's theory, and few physicists nowadays consider this kind of backwards time to be a real possibility. In the contracting phase of the universe, causal influences would propagate backwards from the future, and they would eventually meet other causal influences moving forward in time. Events at the meeting point would then be influenced by both the future *and* the past. Furthermore, it isn't very clear what would happen at the turnaround point. If the universe suddenly began to contract, it would still appear to us that the universe was expanding. We see distant galaxies by the light they emitted millions or billions of years ago. They would not seem to be approaching us until millions or billions of years after the contraction began.

However, physicists have continued to speculate about the possibility of backwards time every now and then. Among them is the British physicist Stephen Hawking. For years, Hawking favored a model of the universe that began with a big bang, expanded to a maximum size, and then began to contract, finally collapsing in a big crunch that was the analogue of the big bang. Hawking also proposed that time would flow backwards in a contracting universe. He didn't base his arguments on the character of heat flow, as Gold had. He suggested instead that the laws of quantum mechanics would cause the arrow of time to reverse.

In 1991, Hawking reversed himself, after discovering that he had made a error in his calculations. He described his backwards-time theory as his "greatest mistake." However, this has not prevented other physicists from considering the possibility of backwards time. For example, physicist James Hartle and Nobel Prize–winning physicist Murray Gell-Mann have suggested that Hawking's mistake can be corrected, and that perhaps time really could run in reverse in the manner that Hawking described.

So might there indeed be a phase of the universe in which time runs backwards? Probably not. In 1998, astronomers observing distant supernova explosions determined that the rate of expansion of our universe is increasing. The universe apparently contains a mysterious *dark energy* that is causing the expansion to accelerate. No one is sure what this dark energy is. Scientists suspect that it may arise from quantum events in empty space. Quantum mechanics implies that subatomic particles are always popping into existence, and then disappearing again in tiny fractions of a second, even in a vacuum. There would be an energy associated with these fluctuations. But no one knows how to calculate it, so identifying it with the dark energy is no more than a plausible guess.

The discovery that the expansion of the universe is proceeding at an ever-increasing rate seems to render the question of what would happen in a contracting universe somewhat moot. However, no one can say with certainty that the acceleration will always continue. It is conceivable, though perhaps not likely, that the phenomena of increasing expansion is only a temporary one. After all, when the universe was only a few billion years old, there was no acceleration. At that time, gravity did act as a retarding force. We can't be perfectly sure that the expansion of the universe will not slow down in the future. Until the nature of the dark energy is better understood, it is only possible to say that the universe will *probably* expand forever.

Whether the universe will always be expanding or not, the fact that there are questions about how the arrow of time would behave in a contracting universe indicates that there is a lot not yet well understood. Scientists would like to be able to say that they know whether backwards-flowing time is possible or not. Unfortunately, they are not yet able to do so.

The Electromagnetic Arrow of Time

Light, radio waves, and other kinds of radiation are called electromagnetic because they are made up of oscillating electric and magnetic fields. The only difference between the different types is wavelength. Radio waves have the longest wavelengths, and gamma rays the shortest, while the wavelengths of visible light are approximately in a middle range.

The behavior of the various different kinds of radiation is described by the laws of electromagnetism, which, like all the other basic laws of physics, are time reversible. These laws predict the existence of both "retarded" waves that travel toward the future, and "advanced" waves that travel toward the past (called advanced because they arrive in advance of their emission). Nevertheless, only retarded waves are observed to exist. Perhaps we should be happy this is the case, since the existence of advanced waves would allow us to send messages into the past, which would create all sorts of causal paradoxes.

Since electromagnetic waves only travel toward the future, we can say that there is an electromagnetic arrow of time. However, the reason why this arrow exists is somewhat mysterious. It is not at all obvious what, exactly, prevents the existence of advanced waves. Naturally, it is possible to ignore this problem and simply accept the fact that only retarded waves exist. Indeed, this is what most physicists do. However, ignoring a problem about time is no way to come to terms with its mysteries.

Perhaps the best place to begin would be to try to see what advanced waves would look like. Suppose that there is a lightbulb in the middle of the room. Ordinary retarded waves travel from the bulb to the walls, where some of them will be absorbed and some reflected. The reflected waves will bounce around the room until they too are absorbed.

If there were an advanced wave, the same thing would happen, except that the waves would be emitted in the future and absorbed in the past. Since the absorption comes first, from our forward-time perspective, it would appear as though the light waves were traveling in the opposite direction, that they were emitted by the walls of the room and finally absorbed by the lightbulb. Whenever there is time-reversed behavior, it will always appear to us that things are moving in the opposite direction.

The behavior of light waves under such circumstances is analogous to the behavior of the ripples that spread outward when a stone is dropped in a pond. The ripples continue moving until they reach the edges of the pond, where they are "absorbed." An advanced wave would be analogous to ripples that originated at the edges and converged upon the middle of the pond where they met a stone that had just been dropped. Upon reaching the stone, the ripples would disappear.

Such inwardly moving waves are not impossible. But they are never seen; they are simply too improbable. In order to produce them, the molecules at the edges would have to vibrate in precisely the right way at precisely the right time. This is as unlikely as seeing a shattered glass come together on the floor and rise up to the edge of a table.

So perhaps the reason that advanced electromagnetic waves do not exist is simply that they are too improbable, as improbable as a large, spontaneous increase in entropy. I say "perhaps" because there is an alternative explanation for the nonexistence of electromagnetic waves traveling backwards in time, one that makes no appeal to probabilities at all.

In 1945 the American physicists John Archibald Wheeler and Richard Feynman published a theory that was based on the assumption that if time-reversed radiation is permitted by the laws of electromagnetism, then it must exist in nature. According to their theory, the reason that we do not see advanced waves is that emission and absorption processes cause them to be canceled out.

The original purpose of the Wheeler-Feynman theory was not to explain the absence of time-reversed waves, but to find a way to solve certain mathematical difficulties in theories that described the interactions between charged particles and electromagnetic fields. Wheeler and Feynman achieved only partial success, but their theory aroused great interest among cosmologists because it suggested that there might be a connection between the electromagnetic arrow of time and the expansion of the universe.

Wheeler and Feynman began by assuming that electromagnetic radiation could be emitted in both time directions. First they considered the case where the ordinary retarded radiation was eventually absorbed by matter of one kind or another. They found that when this happened the particles that made up this matter would re-emit radiation in both directions. Half of it would continue on into the future, but half would travel back into the past.

Thus there would be two time-reversed waves, one that propagated from the original emitter, and one that rebounded backwards from the "absorber in the future." The two physicists calculated that these two waves would exactly cancel each other out. The crests of one wave would line up with the troughs of the other. As a result, only ordinary retarded waves would be seen.

Of course, the process would not be quite as simple as I have made it seem. If Wheeler and Feynman's theory is correct, then there must be many different absorbers at many different times and places in the future. Consequently numerous different waves would be traveling back and forth at many different times. However, when the effects of all the waves are added up, what remains is an ordinary forward-in-time wave that has exactly the intensity it should have.

In order for the theory to work, all of the radiation originally emitted must eventually be absorbed by something. However, it is by no means obvious that this would be the case. The average density of matter in the universe is of the order of one atom per cubic meter. As a result, radiation can travel immense distances without encountering any matter. If this were not the case, we would not be able to see galaxies that were billions of years old. The light they emitted would be absorbed before it got here.

You shouldn't be surprised, by the way, that the matter density of the universe is so low. There are indeed huge quantities of matter concentrated in stars and in galaxies. However, the distances between them are so great that the average density is low. If you put a pea in the middle of a football stadium, and averaged out its mass over the whole stadium, you would get a very low figure, although one that would be much higher than the density of matter in the universe.

If the Wheeler-Feynman theory is to work, all radiation has to be absorbed eventually. It doesn't matter precisely when. If, at some time in the future, the universe began to contract, then this absorption would eventually take place. A contracting universe would eventually become so dense that it would become opaque to radiation emitted in the past.

As we have seen, however, it is not likely that the universe ever will enter into a stage of contraction; the expansion is accelerating. Thus billions of years in the future, the universe will be even less dense than it is now. If this is the case, there must be some radiation that will never be absorbed; it must travel forever through a progressively more empty space. And if there is some radiation that is never absorbed, there must be some advanced waves traveling into the past.

Thus it seems possible that advanced waves exist, but that their intensity is so low that they have never been detected. In 1972 this possibility was tested in an experiment performed by the American astrophysicist Bruce Partridge. Partridge used a transmitter to beam microwaves (short-wavelength radio waves) into space. He pointed the antenna in such a way that the microwaves would avoid the Milky Way, and eventually travel into intergalactic space. He estimated that 3 percent of the waves would be absorbed by the atmosphere and less than 1 percent by the Milky Way galaxy. Thus the bulk of them would travel through space for billions, possibly trillions of years.

Partridge transmitted the microwaves in pulses that were a thousandth of a second long. In the gaps between the pulses, the waves were directed into an absorber attached to his transmitter. This enabled him to compare a situation where all the microwaves were absorbed, with one in which they might be absorbed at some time in the distant future. He reasoned that, if the Wheeler-Feynman theory was correct, and if not all of the beamed microwaves were eventually absorbed, then some backward-propagating waves would be present. From his own forward-time point of view, it would appear that some energy was flowing into his apparatus, because advanced waves would look like waves that were converging on the antenna and being absorbed by it.

If there was some small amount of energy flowing into the apparatus when the microwaves were beamed into space, and no energy flowing into it when they were directed toward the absorber, the power consumption would not be exactly the same in the two cases. The transmitter would be using slightly *less* power under the former circumstance. But Partridge found no difference in the power consumption, to an accuracy of one part in a billion. This seemed to indicate that either the Wheeler-Feynman theory was wrong, or that 100 percent of the waves beamed into space managed to get absorbed by the universe at some time in the future.

As we have seen, it is unlikely they would all be absorbed in an expanding universe. So, unless some future experiment gives a different result, then the Wheeler-Feynman theory should probably be put aside. It appears likely that the reason we do not see advanced waves is *not* that they are canceled out in a complicated way, but simply that they would be too improbable.

It would be nice if the Wheeler-Feynman theory, or some other theory like it, happened to work. It is always more satisfying to explain physical phenomena in terms of the fundamental laws of physics without appealing to probabilities. However, at present, the probabilistic explanation of the electromagnetic arrow of time seems the most plausible. And the electromagnetic arrow appears to be very similar to the arrow defined by the direction of increasing entropy.

It may be that the situation will change in the future, that someone will discover some fundamental reason why electromagnetic waves travel only into the future, and never into the past. If that happens, scientists may also discover that there is some relationship between the electromagnetic arrow and the all of the other four arrows of time. But it would be futile to speculate about what might be discovered in the future. The only thing we can be sure about is that, whatever is discovered, it is likely to be something that will surprise us.

Kaons

Reactions between subatomic particles are generally time reversible. If a reaction can take place, the reverse is possible also. For example, a uranium-238 nucleus will sometimes spontaneously emit an alpha particle (a composite particle made up of two protons and two neutrons), transforming itself in a nucleus of the element thorium in the process. If a thorium nucleus is bombarded with an alpha particle of appropriate energy, the reverse process will take place. The alpha particle will be absorbed, and the thorium will be converted into uranium. Consequently, if we could make a videotape of such an event, no one to whom we showed the tape would be able to tell whether it was being run forward or backwards.

Most reactions between single particles are time reversible, too. For example, if a neutrino (a particle that has a very small mass, much less than that of an electron, and zero charge) collides with a neutron, a proton and an electron can be produced. The reverse process can also take place; the collision of a proton and an electron can produce a neutron and a neutrino. Thus if we were shown a videotape of either process, we again would not know whether the tape was being run forward or backwards.*

There is one known exception to the rule that reactions between particles are time reversible. This is the decay of the neutral kaon. Kaons are short-lived particles that that are produced in collisions between such particles as protons and neutrons. They belong to a class called the mesons; in fact, kaons were originally called K mesons. There are many other kinds of mesons. All of them are unstable, and are seen only in experiments performed in the laboratory.

There are three varieties of kaon. The particle can have either either a positive or a negative charge, or no charge at all. The positively and negatively charged kaons exhibit perfectly normal time-reversible behavior. Neutral kaons do not. When neutral kaons decay, they can decay either into three particles; for example, three pions (a pion is another meson), or into two. A decay into two pions takes about a ten-trillionth of a second, while a decay into two takes thousands of times longer.

Kaons decay into three particles most of the time. Originally it was thought that this always happened. But in 1964, it was discovered that the decay into two particles happened on rare occasions. This was a surprise, because theoretical calculations indicated that a decay into two particles was not time reversible. On some occasions, kaons behaved in such a way as to exhibit a sensitivity to the direction of time.

In 1998 the time-asymmetry of the decay into two particles was

---

*In reality, no such videotape could be made. These particles are too small to be seen, even by powerful electron microscopes. However, in "thought experiments" like this one, all things are possible.

observed directly, by teams of physicists at Fermilab (Fermi National Accelerator Laboratory) in Batavia, Illinois, and at CERN (Centre Européen pour la Récherche Nucléaire) in Geneva, Switzerland. Thus it was confirmed experimentally that a rare kaon decay did indeed define an arrow of time.

The story does not end there. In 2001, it was confirmed that another meson, called the B meson, also exhibited time-asymmetric behavior. Two teams of physicists, one at the Stanford Linear Accelerator Laboratory (SLAC) in California and at KEK, a Japanese particle accelerator laboratory in Tsukub, studied the ways in which this meson decayed and found that some of the decays were sensitive to the direction of time. As in the case of the kaon, the effect was small. It was, however, measurable.

So far, no time asymmetry has been seen in the particles that make up ordinary matter: the proton, neutron, and electron. Some physicists think that the neutron might exhibit a very small time asymmetry. However, the most recent measurements indicate that, if this effect does exist, it must be very, very small, too small to observe with existing experimental technology. There is also a possibility that atoms and molecules might sometimes exhibit time-asymmetric behavior. But when I wrote this chapter in 2001, this had not yet been observed (and if no footnote has been added here, it will mean that it had still not been observed by the time this book went into galleys).

There remain a lot of unanswered questions about the arrows of time. But, as we have seen, questions concerning time asymmetry are being actively researched. No final conclusions have been reached. However, it is possible that new discoveries about the time-asymmetric phenomena may compel physicists to modify some of their theories about the behavior of elementary particles. This, in turn, may lead to new knowledge about the creation of, and the fundamental properties of, our universe. For example, there is the question of why time asymmetry exists at all. A universe in which *all* interactions between particles is time reversible can certainly be conceived. This raises the questions of why our universe is not like that, and of how it is different from a hypothetical time-symmetric universe.

Alternatively, the discovery of new theories about the nature and properties of matter may lead to a new understanding of time. The British physicist Roger Penrose has suggested that gravity might have something to do with the time-asymmetric behavior of the neutral kaon. In particular, he believes that a theory of quantum gravity, one that combined Einstein's theory of gravity—his general theory of relativity— with quantum mechanics would explain why there are arrows of time. Penrose admits that he doesn't know how this would be explained, so his idea isn't much more than a hunch. However, it an intriguing one. So far, all attempts to create a theory of quantum gravity have failed. Although quantum mechanics and general relativity are both extremely successful and well-confirmed theories, there are ways in which they are incompatible with each other. And no one has found a way to overcome this incompatibility. However, there has to be some reason why time asymmetries exist, and Penrose could possibly be on the right track.

But perhaps I shouldn't try to speculate in this vein too far. Talking about what theories *might* be discovered in the future, and about what they might possibly tell us isn't likely to be a way to reach any firm con- clusions. What physicists can be certain about is that the kaon, a particle that is seen only in the laboratory, exhibits behavior that is different from that of other known particles. There are theoretical reasons for believing that another particle, the B meson, will be found to behave similarly. These particles are somehow "aware" of the direction of time, while most other particles are not. If physicists eventually discover why this is the case, they will have penetrated one of time's deep mysteries.

What Is Time?

There is much more than can be said about time. Indeed, entire books have been written on the subject. In this chapter, I have chosen to focus on one of the most mysterious aspects of time: the existence of five dis- tinct "arrows." As we have seen, there are numerous unanswered ques- tions about time asymmetry. No one really knows how the five arrows of time are related to one another. It is not known whether, under some circumstances, time could go backwards, either in the entire universe, or in some portion of it. We don't really understand why all the known fundamental laws of physics should be oblivious to the direction of time, or why a small number of particles exhibit time-asymmetric behavior, while others do not. Time-asymmetric behavior has only been observed in short-lived particles. Some physicists hope to be able to observe it in such stable objects as atoms and molecules. However, they have not succeeded.

Furthermore, the relationship between physics time and subjective, psychological time remains a mystery. Physics knows nothing of the idea of "now" that is so important to us in our daily lives. Time enters the equations of physics in the form of intervals between events. A particle may be created, and decay a trillionth of a second later. Approximately 10 billion years may elapse between the birth and the death of a medium-size star like our sun. Radio waves may oscillate at the rate of so many cycles per second. A pendulum will take so many seconds to complete one of its swings. There isn't any "now" in any of those phenomena, unless some individual, usually a scientist, happens to observe one of them at some particular time. But then it is his "now" that comes into play.

We do not even know why time exists. It is possible to conceive of universes that do not have a dimension of time, and, for all we know there may be parallel, timeless universes. In fact, according to a hypothesis proposed by Stephen Hawking and James Hartle that was expounded in Hawking's book, *A Brief History of Time*, our universe may have begun in such a timeless state. According to Hawking and Hartle, time originally resembled a dimension of space (this is what Hawking means when he speaks of "imaginary time"), so that initially there were four spacelike dimensions, and nothing resembling time as we know it. According to their hypothesis, the universe had no beginning in time. Time came into existence only when the spacelike imaginary time evolved into the dimension of time that we know.

It is not my purpose here to expound the Hawking-Hartle hypothesis in detail. The only point I want to make is that it is relatively easy to imagine conditions under which there is no time dimension. It would be difficult to conceive of a world with more than one dimension of time. No one knows what a second time dimension could possibly be. But picturing a universe in which there is no time is relatively easy.

There are a lot of unsolved mysteries about time. To be sure, some progress is being made. The study of time asymmetry in the behavior of certain particles is a promising line of research. Surprising new discoveries are likely to be made in the years ahead. However, at the moment, science is in a position not unlike that of St. Augustine, who knew what time was until someone asked him to explain it.

2 | *Does the Future Already Exist?*

Suppose someone says to you that the NASDAQ index will go up tomorrow. You answer, "No, it won't." Since the predictions the two of you have made are mutually contradictory, one must be true and one must be false. Obviously, the index cannot both rise and fall. But if one of the statements is true, then it follows that the future is already determined. If one of the statements is true today, then it must be true tomorrow.

You probably won't find this argument very convincing. A multitude of unpredictable factors can influence stock prices. No one knows in advance what the psychology of the market is going to be or what unexpected events are going to happen. But that doesn't really refute the argument. It is true that neither of us can predict the future behavior of the stock market, but that doesn't alter the fact that one of us made a statement that was true, while the other made a statement that was false.

Aristotle presents this argument in chapter 9 of *On Interpretation*, and gives this example: Either there will be a sea battle tomorrow, or there will not. Aristotle did not think the argument for determinism that supposedly followed from this was a good one. And most subsequent philosophers have not thought so, either. However, they have generally found that it is not so easy to say precisely what is wrong with the argument. As a result, the passage in which Aristotle presented the problem has become the most discussed and argued about in philosophy. Abelard, St. Anselm, St. Thomas Aquinas, Aristotle, St. Augustine, Avicenna, and Averroes are among those who tackled the problem. And those are only some of the *A*'s. The problem was extensively discussed in medieval times, and it is still being cited today. According to the British philosopher J. R. Lucas, "Almost any solution that has been offered is open to some fatal objection: and almost every solution has been offered."

Lucas doesn't claim to be able to solve the problem. "I am not going to try to unravel all the arguments that go to create Aristotle's problem about tomorrow's sea battle," he says. "I have fared no better in my attempts than any of my medieval predecessors." Naturally, I am not audacious enough to make an attempt, either. My only purpose in introducing the argument is to show that understanding the character of the future is not so simple a task. The past is generally taken to be fixed and unalterable. We generally believe that the future has a different character. But does it? Is it possible that future events are determined by events in the present?

Questions of the character of the past and future arise in connection with some of the things I discussed about time in the last chapter. For example, suppose that, contrary to what most scientists currently believe, the universe will eventually enter into a phase of contraction, and that time does indeed go backwards in a contracting universe. If the contracting phase of the universe is inhabited by intelligent beings who view time in the reverse sense, then their past is our future. And naturally the reverse is also the case. One view of the direction of time would not be better than the other. Both would be equally valid. Our big bang would be their big crunch, and what we call the big crunch would be seen by them as a big bang. Such a universe would have no end. Instead it would have two beginnings.

Under such circumstances, the future—or at least the part of the future that constituted their past—would have to be regarded as unalterable, like our past. The only alternative to this point of view would be to say that both past and future are somehow undetermined. And it is not very clear what it would mean to say that the past is not unalterable.

There probably isn't any point in pursuing this line of thought any further. After all, backwards time is an unlikely theoretical possibility. It is more interesting to take a brief look at some philosophical ideas about past and future and at the concept of determinism, and then discuss what science can tell us about these things.

The Appeal of Determinism

There must be something appealing about the idea of determinism. Deterministic philosophies have flourished in all ages. Even today, when philosophers no longer try to create metaphysical systems, deterministic ideas can be seen in popular culture. Astrology is deterministic, for example. It is based on the idea that events in our lives are somehow correlated with the movements of the planets. Sometimes determinism is seen in an extreme form called fatalism. Soldiers fighting in wars have often spoken of "a bullet with your name on it." Of course, this is never meant literally. It is the expression of the belief that if one is destined to die at some particular time, that fate cannot be avoided.

Most of the individuals who have believed in determinism have nonetheless behaved as though they had free will, and that their actions had consequences for the future. Thus a belief in determinism generally involves one in contradictions. But then, perhaps there is nothing so surprising about that. Human beings are not always perfectly rational, and most of us entertain some sets of beliefs that are mutually contradictory.

One of the first deterministic philosophies in ancient Greece was the atomism of Leucippus and Democritus. Not much is known about the former, except that he flourished around 440 B.C. and was from Miletus. A great deal more is known about Democritus, who was a contemporary of Socrates, so accounts of atomism are generally based on the doctrines he propounded.

The atomists believed that everything was composed of atoms, which were indestructible and too small to be seen by the human eye. They thought there were an infinite number of atoms with numerous different sizes and shapes. These atoms were always in motion, and they often collided with one another like billiard balls. According to Aristotle, the atomists believed that atoms could contain differing quantities of heat; spherical atoms, which were the constituents of fire, were the hottest.

In antiquity, it was sometimes said that the atomists attributed everything to chance. But that only demonstrates that their doctrines were not always well understood. The atomists were strict determinists who held that the motion of atoms, which was governed by natural laws, was the cause of everything that happened. Their world was a world without purpose, a world that operated like some giant machine. And the mind operated according to mechanistic laws, too, they said; after all, it also was composed of atoms.

The atomistic theory took on a somewhat different form in the thought of Epicurus, who was born in 342 or 341 B.C., and who founded a school of philosophy in Athens. Epicurus wanted to allow for free will, so he proposed that the atoms of the mind behaved somewhat differently than those which made up ordinary matter. The former did not always behave mechanistically; free will caused them to sometimes swerve from their paths. However, Epicurus retained the other essentials of atomistic philosophy.

Of all the Greek philosophies, atomism seems closest to modern science. However, the idea of atoms was originally proposed by Leucippus and Democritus for philosophical reasons, not scientific ones. At the time there was no empirical evidence for atoms. The atomists were apparently trying to find some sort of mean between the philosophy of Parmenides, an older contemporary of Socrates, who had maintained that the world was really an unchanging "One," and the more pluralistic ideas advocated by other philosophers. More than two millennia were to pass before atomism became a scientific theory. Even at the beginning of the twentieth century there were still some scientists who doubted the existence of atoms. They viewed atoms as a useful fiction that corresponded to nothing in reality.

Atomism was criticized by Aristotle, and Aristotle's authority finally prevailed. However, it is doubtful that this impeded the eventual rise of science in any way. The first modern scientists, such as Copernicus, Kepler, and Galileo, concerned themselves with astronomy and with the laws of motion, not with the ultimate nature of matter. If atomistic ideas had been common in their day, it is unlikely that this would have had much of an effect on their achievements.

The Past Is the Future

We think of time as something that is linear. It extends into the distant past, back to the moment of the big bang, and also into the distant future. This point of view seems perfectly natural to us, so natural that we find it difficult to conceive how time could be viewed in any other way. However, time was frequently viewed differently in ancient times. For example, in classical Greece it was often thought of as something that was circular. It was believed that every event that ever took place would be repeated endlessly in a series of cosmic cycles. For example, there would be another Trojan War, another Athens, another Socrates, and another drinking of the hemlock. Each of these things had happened innumerable times in the past, and would be repeated innumerable times in the future.

A reference to this idea can be found in Plato's dialogue *Parmenides*. There is a passage in which Parmenides states that something that is getting older must simultaneously be getting younger. The idea is that the thing becoming older is simultaneously moving away from and getting nearer to its beginning in circular time. In the dialogue, Parmenides does not elaborate upon the idea. Plato apparently assumed that his readers would understand perfectly well what was meant.

This idea later became a prominent part of the philosophical doctrine of Stoicism, which was first propounded by Zeno of Citium (no relation to Zeno of Elea, who proposed the paradox of Achilles and the Tortoise) around 300 B.C. The concept of circular time was naturally not original with Zeno. He seems to have regarded it as part of "the wisdom of the ages." Zeno, incidentally, was a contemporary of Epicurus and the philosophies of both men were later to become popular among the Romans. The Romans frequently looked to philosophy rather than religion as a guide to the conduct of life. The philosophies of both men were expounded in a book by Cicero, who lived during the first century B.C., in a book titled *The Nature of the Gods.* Apparently the Romans found deterministic philosophies quite appealing.

According to Zeno the same events were destined to repeat themselves in endlessly recurring cycles. At the end of each cycle, the entire cosmos would be destroyed in a great conflagration, and then it would be created anew. In each succeeding cycle, exactly the same events would recur, and human beings would be powerless to alter them. According to Zeno, the wise man should accept this fact and cultivate an inner virtue; he was powerless to do anything else.

You might think that the eternal recurrence posited by the Stoic philosophers bears no resemblance to anything encountered in modern science. But this is not entirely true. Near the end of the nineteenth century, the German physicist Ludwig Boltzmann suggested that the universe might indeed go through recurring cycles over enormous periods of time.

Boltzmann was one of the founders of a branch of physics called statistical mechanics, which deals with the behavior of large ensembles of molecules; for example, the molecules that make up a gas in a container. During the 1870s Boltzmann published a series of papers in which he showed that the law of increasing entropy could be explained in terms of the statistical behavior of molecules. Boltzmann thought he had proved that entropy must always increase as the result of molecular collisions. If this were true, then the law of increasing entropy would have been a direct consequence of Newton's laws of motion.

But the French mathematician Henri Poincaré soon found a hole in Boltzmann's argument. Poincaré proved that a collection of molecules enclosed in a container must always return to a previous state after a long period of time. Thus, if the molecules had initially been in a state of low entropy, then however much the entropy increased, the initial state would eventually have to recur. The time that this would take was enormous. However, Poincaré's calculation was mathematically rigorous. There would be recurrence after some sufficiently long period of time.

Boltzmann wasn't happy about this conclusion, and he struggled to find some way to avoid it. Yet all his efforts were unsuccessful; he could find no way around Poincaré's proof. In the end, he suggested that this result could be applied to the entire universe. If the universe was finite, and if it was previously in some low-entropy state, then that state would eventually have to recur. There would eventually be an enormous fluctuation that would transform a high-entropy universe into one of low entropy.

If time were endless, then this would happen innumerable times. To be sure, the length of the cycles would be enormous. A time equal to trillions of times the present age of the universe would not be sufficient for a recurrence to occur in a single liter of air. Nevertheless, if one assumed that future time was infinite, then there would necessarily have to be one recurrence after another.

Such recurrence would not be the same as the eternal recurrence of the Stoics. As we have seen, the Stoics were determinists. Succeeding cycles of the universe would not be much like the one before. Every time the universe reached a state of low entropy, it would evolve from that state in a random way. But this didn't deter the German philosopher Friedrich Nietzsche, who pointed out that if time was infinite, then any given configuration of atoms on Earth would be repeated innumerable times in the future. The same events would happen again and again and again.

We now realize the huge flaw in these arguments. They only work if the universe is static. But the universe is not static; it is expanding. If the expansion continues, then no previous state can ever recur. Naturally, this never occurred to Boltzmann or to Nietzsche, who assumed that the universe must be stable. It had been believed since the time of Aristotle that the universe had a constant size, and this idea had never even been questioned. Even Einstein made this assumption when he began to work out the consequences of his law of gravity, the general theory of relativity. It wasn't until Hubble discovered the expansion of the universe that Einstein discovered he was wrong.

Determinism in Newtonian Mechanics

In 1705, some eighteen years after the publication of Newton's *Principia*, his great work on mechanics and on gravitation, the English astronomer Edmond Halley calculated that the comets that had been seen in 1531, 1607, and 1682 were really one comet and predicted that it would return again in 1758. When it did make an appearance that year, the comet was named after Halley, and it has been known as Halley's comet ever since.

Halley had been able to make this prediction because he had used Newton's law of gravitation and his laws of motion to compute the orbits of some twenty-four comets. He was able to do this because Newton's laws were rigidly deterministic. The past behavior of astronomical objects could be used to predict their behavior in the future.

Today, Newtonian mechanics (mechanics is the branch of physics that deals with forces and energy and their effects on bodies) can be used to predict the behavior of moving bodies with far greater accuracy. It can be used to predict the path of projectiles that are fired from the surface of the earth, the paths of space vehicles. The orbits of the planets in the solar system can be predicted for thousands of years in the future.

Before quantum mechanics was discovered, it was thought that Newtonian mechanics could be used to predict the future behavior of any collection of particles. If one knew the initial positions and velocities of the particles and their masses, then it would be possible to predict their motion for all time, at least in principle. To be sure, there were numerous cases in which the practical difficulties would prevent this. For example, it would hardly be possible to determine the position and velocity of each molecule of a gas in order to predict the future behavior of the gas. If this could be done, no existing computer would be powerful enough to compute the future collisions that would be experienced by each one. But this is only the result of our limited knowledge and computing ability. If Newtonian mechanics was always valid, then the future behavior of each particle would be rigidly determined. Everything that would happen in the future would be a consequence of the conditions that existed in the past.

The determinism of Newtonian mechanics is perhaps best summed up in an often-quoted statement made by the late-eighteenth–early-nineteenth-century French physicist Pierre-Simon de Laplace. According to Laplace,

An intelligence knowing, at any given instant of time, all the forces acting in nature, as well as the momentary positions of all things of which the universe consists, would be able to comprehend the motions of the largest bodies of the world and those of the smallest atoms in one single formula, provided it were sufficiently powerful to subject all data to analysis; to it, nothing would be uncertain, both future and past would be present before its eyes.

Laplace knew very well that such a calculation could never be performed. However, that fact apparently did not negate his conclusion, that the future behavior of every particle in the universe was already determined for all eternity.

Newtonian mechanics conceived of the universe as a kind of gigantic clockwork mechanism. Not only were all future happenings already determined, there seemed to be no room for free will. Our brains, after all, are made up of material particles, too. If their behavior is determined, then any brain state must be the consequence of previous brain states. There is no room left for human freedom. We can do nothing to alter a future that is already fixed.

Laplace's argument seems to be much less convincing today than it appeared to be in his time. It is now known that the behavior of many physical systems is chaotic. This means that tiny changes in the initial conditions can have very large consequences. As a result, the behavior of chaotic systems is unpredictable. The earth's weather patterns, for example, appear to be chaotic. It is often impossible to make accurate forecasts more than about five days in advance, and on occasion the next-day forecasts turn out to be terribly inaccurate. This is not a result of our inability to measure the factors that influence the weather accurately. It is a result of the fact that small perturbations can have very large effects. For example, it has been said that the fluttering of a butterfly's wings in the Amazon can cause a hurricane in the Atlantic. Perhaps this statement should not be taken too literally. The weather is subject to a large number of influences, and the flight of the butterfly is just one. However, tiny perturbations in weather patterns can grow rapidly. And since we never know what these small perturbations are, it is not likely that we will ever be able to forecast the weather to the accuracy that we would like.

Chaotic systems are still deterministic, however. The discovery of chaos did not prove that Newtonian mechanics was not deterministic. It only implied that there exist systems whose future behavior cannot be predicted unless we know the initial conditions to almost-infinite accuracy. But there is another, possibly stronger objection that can be made to Laplace's argument. In his day, it was believed that the laws of Newtonian mechanics were exact. Today, we know this is not true. We realize that all known physical laws are only approximations. For example, Newton's laws have been superseded by those of relativity. Under most circumstances, relativistic corrections to the predictions of Newtonian mechanics are tiny. They are so small that the differences between the predictions of Newtonian mechanics and of relativity cannot be measured. No scientist would use relativity to calculate the future motion of the planet Saturn, or the path of a space vehicle to one of the moons of Jupiter. Using Newton's simpler law of gravitation gives results that are accurate to a very large number of decimal places.

There is no reason to think that the predictions of relativity are exact, either. In fact, physicists believe that the laws of relativity *must* break down under extreme conditions, such as those that exist near the center of a black hole. At best, relativity is only approximate. It is even possible that there are no exact laws of nature. It is conceivable that, at best, we will only be able to discover more and more accurate laws, that progress in physics will always be something like peeling away the numerous layers of an onion. In other words, it is possible we will never know everything.

But if Laplace's argument is not entirely convincing, it still retains some of its force. It at least seems to suggest that although the future may not be completely determined, it is still determined to a high degree of accuracy. If Laplace's demon ("Laplace demon" is the name given to the "intelligence" he speaks of) were somehow able to use Newtonian mechanics to predict that next week someone would fall into the path of an oncoming subway train, it wouldn't make much difference if his calculation of the place where he would land on the tracks contained an error of a thousandth of an inch, or even one or two feet.

The final result would be the same. Of course, I am assuming this person's behavior is not chaotic. If it is, the demon might be able to make the prediction only a few seconds in advance.

The Indeterminacy of Quantum Mechanics

There are many things in quantum mechanics—for example, the energy levels of a hydrogen atom—that can be calculated exactly. There are many other quantities that can be approximately computed to a high degree of precision. However, quantum mechanics cannot be used to calculate when an event is going to take place. For example, a free neutron—a neutron not bound in an atomic nucleus—will, on average, decay into a proton, an electron, and a neutrino in about fifteen minutes. However, quantum mechanics cannot tell us when any given neutron will decay. If we have a large number of free neutrons, their decay times will seem totally random.

Similarly, a television picture is created by large numbers of electrons striking a fluorescent screen. Every time an electron strikes the screen a pinpoint of light is created. It is possible to determine what the average behavior of these electrons will be. If this were not the case, we could not manufacture television sets that produced pictures of any kind. However, if an individual electron is set in motion toward a fluorescent screen, it is not possible to say exactly when and where it will strike. Quantum randomness prevents us from knowing its trajectory in advance. Television sets exist only because the random fluctuations average out when there are very large numbers of particles.

A television set provides a good example of the fact that quantum randomness is generally observed only on the subatomic scale. There is quantum randomness everywhere in the world, yet quantum effects are very small. They are large enough to profoundly affect the behavior of particles such as electrons and neutrons, but far too small to have any measurable effect on most macroscopic objects. Quantum randomness does not affect the motion of a billiard ball, or the orbits of the planets around the sun, or the functioning of a refrigerator.

It is the indeterminism of quantum mechanics that finally kills off the Laplace demon. Quantum randomness doesn't alter the fact that we are still able to predict some quantities, such as the future position of the earth, many thousands of years in the future. However, it does destroy the idea that *everything* is determined in advance. Matter, after all, is made of subatomic particles. It isn't possible to say that something must behave deterministically when its components do not.

As we have seen, Epicurus speculated that there was a connection between free will and the sometimes random behavior of the atoms that made up the mind some 2,300 years ago. Quantum mechanics tells us that the atoms that make up the brain must indeed sometimes behave in random ways. This suggests that there might be some connection between human freedom and quantum indeterminacy.

It is sometimes said that the British astronomer Sir Arthur Eddington made such a suggestion in the early 1930s. I'm not sure how this story arose; in fact, Eddington branded the idea as "nonsense." In Eddington's opinion, the assumption of free will implied that the brain must contain "conscious matter" that correlated processes taking place in the remainder of the nervous system. If free will did operate, Eddington said, it did so without regard to quantum indeterminacies. At best, chance events on the subatomic level could introduce chance elements into human behavior. But chance elements would not make the will free; they had nothing to do with the idea of free choice.

Eddington's argument was a philosophical not a scientific one. His phrase "conscious matter" was nothing more than a novel name for "mind." But Eddington realized that his speculations were philosophical in nature. Indeed, he spoke of consciousness as something that was "outside physics." When he introduced the notion of "conscious matter" he realized that he was talking about something that had never been seen in a laboratory.*

Since Eddington's arguments were philosophical ones, it might be worthwhile to examine the question of what effects quantum randomness might have on the functioning of the brain in more detail. I think we will be able to see that if there are indeed any effects, they are prob-

---

*Scientists today understand much more about the mind than they did in Eddington's day. However, they speak of free will only rarely, and a discussion of their discoveries can safely be left for a later chapter.

ably not significant ones. A neuron is much too large an object to be affected by quantum events in any significant way. In fact, the proteins of which a neuron is made are themselves large, compared to individual particles. Proteins are very complex molecules that interact with one another because they have characteristic shapes. Proteins act on one another because these shapes cause one protein to fit into another. For example, a protein can act as a catalyst. If two other proteins fit into a large protein, then this can cause the two smaller protein molecules to be joined together. The interatomic forces that cause the proteins to have the shapes that they do can be explained by quantum mechanics, at least in principle (in practice, this would be a problem far too complex to solve). However, it is not likely that quantum randomness plays a very significant role.

But suppose that quantum indeterminacy does indeed affect the functioning of neurons in some way we don't understand. Even in this case, it is hard to see how there could be noticeable effects on mental events. The brain appears to be "wired" together in such a way that chance happenings will not affect it much. Connections between neurons are set up in such a way that there is a great deal of redundancy. A nerve impulse can get from one section of the brain to another along many different paths. If some chance event interrupts the impulse along one of these pathways, it will continue to be propagated along many others.

Every day, approximately 10,000 of our brain cells die, and are not replaced. The brain's neural redundancy ensures that these cell deaths will not have any significant effect. And if the daily disappearance of large numbers of neurons has no significant effect on mental functioning, it is hard to imagine that much smaller quantum fluctuations would.

Quantum events could conceivably affect the firing of some of the neurons. But, in view of the brain's redundancy, it is not obvious that this would do anything more than create a small amount of "noise" or "static." Brains are not constructed like electronic devices, which may behave oddly, or fail altogether if one component begins to behave erratically. Thus we should probably conclude that quantum indeterminacy and free will probably not have anything to do with each other.

But it is impossible to prove a negative. Thus it is still possible to engage in speculation, and indeed some scientists have. For example, Roger Penrose has suggested that consciousness—and presumably also free will—might have something to do with quantum effects in small components of cells called microtubules.

Penrose's ideas are highly speculative. Not only do they depend on the existence of quantum effects that have never been observed, they also depend on certain assumptions that Penrose makes about a hypothetical theory that might someday supersede quantum mechanics. Since Penrose's ideas are not based on any empirical evidence, they are essentially philosophical in nature. He is, of course, a scientist, and he makes use of scientific ideas. However, he advances no testable scientific hypotheses. Or at least there is nothing in his theory that can be tested at present.

Back to Square One?

It appears that science is no more able to provide us with any final answers to questions about determinism and free will than philosophy is. As we have seen, Laplace's argument for determinism is fatally flawed. However, it does not follow that the behavior of macroscopic objects is not largely deterministic. Proving that an argument for an idea is wrong does not necessarily prove that the idea is incorrect. If I use an invalid argument to "prove" that $2 + 2 = 4$, it doesn't follow that $2 + 2$ must equal 5.

Quantum mechanics tells us that the behavior of subatomic particles is indeterministic, but quantum effects do not generally affect the behavior of large objects in measurable ways. Furthermore, we cannot be entirely sure that quantum mechanics will not eventually be superseded by a theory in which determinism is restored. Einstein, for one, believed that this would eventually happen. Today, most physicists do not think this is very likely. Quantum mechanics is one of the most successful and best-confirmed theories that science has ever known, and scientists believe that it tells us something fundamental about nature on the subatomic level. There are good reasons to believe that if quantum mechanics is eventually superseded by a deeper theory, then the indeterminism will remain.

Free will has always been a problem for philosophers. We don't know why human freedom exists. We can't even be sure that it is not an illusion. Science appears unable to provide any answers to such questions. We certainly can't draw any firm conclusions about free will from quantum mechanical indeterminism. In chapter 11 I will be discussing some discoveries that have been made about mental functioning, but I will have little or nothing to say abut the free will question. Cognitive scientists rarely even mention it; no one really knows how it could be approached.

On the other hand, it wouldn't be accurate to say that science tells us nothing about determinism and free will. Even though no definite conclusions have been reached, the raising of these questions has at least delineated some of the scientific issues involved. And, as we shall see, there does seem to be a little more to be said about the subject. There is an interpretation of quantum mechanics that seems to put matters in an entirely different light.

The Strange World of Quantum Mechanics

Quantum mechanics is an extremely well-verified theory that can be used to explain a wide variety of phenomena. Some of its predictions have been experimentally confirmed to an accuracy of better than 1 part in 10 billion. When viewed solely as a tool for calculation, there seems to be nothing very problematical about it. Indeed, this is how most physicists look at quantum mechanics. It is something they can use to perform calculations and it does the job superbly.

But as soon as one begins to ask what quantum mechanics "really means," problems arise. The picture of subatomic reality it gives us is a strange one, and it can be interpreted in a number of different ways. The problem is that, if quantum mechanics is correct—and there is every reason to think it is—then a particle can be in a number of different states at the same time. But when the particle is observed, one of these states becomes real, while the others disappear.

For example, early in the twentieth century, before quantum mechanics was discovered, it was imagined that the electrons in a atom followed precisely defined orbits around the nucleus that were similar to the orbit of a planet circling the sun. At any point in time, an electron had a definite location. Scientists might not have been able to determine exactly where an electron was; individual electrons were too small to see, even in the most powerful microscopes. But no one doubted that this picture was accurate.

Quantum mechanics changed all that. It became necessary to conclude that the position of an electron could only be described in terms of probabilities. The electron could be anywhere—and everywhere—at any moment of time. The picture of an electron as an orbiting particle was discarded, and replaced by the idea of an electron cloud. Quantum mechanics required that the position of the electron had to be "smeared out" in a pattern that surrounded the nucleus.

Electrons that are not in atoms exhibit a similar kind of behavior. Their position can generally be described as clouds of probabilities. And yet when an electron is observed, for example, by allowing it to strike a fluorescent screen, it will always appear in some particular place.

You might think it would be simpler to assume that electrons and other particles do follow definite paths, and that we simply don't know what these paths are. Yes, it would be simpler, but it has been proved experimentally that this interpretation is incorrect. For example, it is possible to cause a particle such as a neutron or a photon (a photon is a particle of light; according to quantum mechanics, light has some particle characteristics, and particles sometimes behave as waves) to follow two different paths simultaneously. This has been observed experimentally. The particle can be made to follow both paths, and it can then interfere with itself when the two paths are brought together again. But this only works if no attempt is made to observe the particle as it travels from one place to another. If an observation is made (remember that it is the observation that causes probabilities to dissolve and become real), then the particle turns out to be following either one path or the other, and there is no interference.

There are some especially simple examples that can be used to clarify the implications of quantum mechanics. One of these depends on the idea of "spin," which is a characteristic of all of the particles of matter.

Quantum mechanical spin is a relativistic effect, and it isn't quite the same thing as the spin of a macroscopic object. There are some similarities, however, and it does no great harm to picture an electron or some other particle as a tiny rotating object.

When the spin of an electron is measured experimentally, in any direction, there are only two possible results, called "spin up" and "spin down." Here "up" and "down" are just convenient labels used to designate clockwise and counterclockwise motion. According to the standard interpretation of quantum mechanics, it is *not* correct to say that the electron has a spin that is either up *or* down and that we discover which spin it had when we make the measurement. It is necessary to conclude that the particle had a spin that is 50 percent up and 50 percent down before it is observed, and that it assumed one spin state or the other when we "looked" at it (the spin state of a particle can be measured even though the particle cannot be seen in a microscope).

The behavior of quantum objects can be interpreted in numerous different ways, and in the enormous literature on the subject scientists and philosophers have expressed various different opinions about the nature of quantum reality. Indeed, there are questions that have not satisfactorily been answered. For example, at exactly what point does the state of 50 percent up/50 percent down collapse into one of the two different possibilities? Does this happen when the electron interacts with the measuring apparatus? Does it happen when the result enters into the consciousness of an observer? Does it happen at some other point in time? Does a quantity such as the spin of an electron have any objective reality before a measurement is made? Can an observation made in the present affect what path a particle followed in the past? In a certain sense these are philosophical questions rather than scientific ones. As I pointed out previously, quantum mechanical calculations yield the same results whatever interpretation one adopts. They are questions about the "meaning" of subatomic reality, not questions about the numerical predictions that quantum mechanics makes. But that only makes the questions all the more puzzling.

The Many Worlds Interpretation of Quantum Mechanics

In the early 1950s, Hugh Everett III, a graduate student at Princeton University found himself puzzled by some of the more common interpretations of quantum mechanics.* He began to wonder what kind of picture would emerge if one took quantum mechanics literally. What would happen if one assumed that a particle was simultaneously in two different states both before and after a measurement? What would happen if one considered both possible outcomes of an experiment to be equally real?

Such questions led Everett to work out what is known as the "many worlds interpretation" of quantum mechanics, which was published in the journal *Reviews of Modern Physics* in 1957, with an accompanying paper by Everett's faculty adviser, John Archibald Wheeler. According to Everett, every possible quantum choice was realized in one possible universe or another. For example, when the spin of an electron was measured, the universe split into two parallel universes, one in which the electron had spin up, and one in which it had spin down. In any other case in which there were two or more quantum possibilities, the universe would split again.

If Everett's interpretation was correct there had to be a staggeringly large number, perhaps an infinity, of parallel universes. Furthermore, additional splittings would have to take place at numerous different moments of time in the future. It wasn't necessary to perform any experiments to make the universes split. That would happen whenever a quantum mechanical "choice" was made—for example, when an atom either emitted or didn't emit a photon of light, when an electron followed one of two possible paths, when a radioactive nucleus decayed (or didn't decay), and so on.

Every time the universe split, the individuals in it would split, too. According to the theory, numerous identical copies of each of us were being created at every moment, and numerous identical copies of each of us had been created at numerous different times in the past. According

---

*I'll have more to say about these interpretations in the following chapter.

to the theory, the "parallel worlds" of science fiction corresponded to something real. Some of the words differed from ours only in very small ways; for example, a glass might be positioned in a slightly different place on a table. Others would be very different. There would be worlds in which the South won the Civil War, and there would be worlds in which the Roman Empire still dominated the world.

It sounded fantastic, but Everett's interpretation was perfectly consistent with every experiment involving quantum mechanics that had ever been performed. Like the other interpretations, it was only a picture of reality. Nothing in it changed the way that quantum mechanical calculations were done. However, even supporters of the interpretation sometimes took pause. Physicist Bryce DeWitt, who popularized the interpretation, wrote of the shock he had felt when he first encountered the "idea of 10^{100} slightly imperfect copies of oneself constantly splitting into further copies" (here 10^{100} is the number represented by the numeral *1* followed by 100 zeros).

Today the leading champion of the interpretation is British physicist David Deutsch, who has attempted to make the idea more palatable by doing away with the idea of universes splitting into copies of themselves. According to Deutsch, it is more reasonable to assume that an infinite number of universes have always existed, and that when a quantum "choice" is made, they simply partition themselves into two groups, one in which one quantum alternative is realized, and one in which the other alternative happens. In Deutsch's version of the idea, universes never split. However, they do sometimes fuse together. For example, when a particle simultaneously follows two different paths and then interferes with itself, what is really happening is this: the particle follows each of the paths in two distinct universes, and when the interference takes place, these two universes come together, making one reality.

If one accepts this interpretation, then it is necessary to conclude that entire parallel universes, containing all the stars and galaxies, exist at the same time and, as Deutsch puts it, "in a certain sense in the same space." Since these universes normally do not communicate with each other, we remain oblivious to their existence. It would not be possible to travel from one universe to another, or to communicate with another universe "in any large-scale way."

Deutsch emphasizes "large-scale" communication because he believes that, even though we could not carry on conversations with the people in other universes, it should nevertheless be possible to perform an experiment by which the existence of these universes could be confirmed. Deutsch's proposed experiment cannot be performed today because it requires a quantum computer, a computer that would make use of quantum processes to carry out calculations. Only this kind of device, Deutsch thinks, would be sensitive enough to catch a glimpse of some other universe's existence. It will not be possible to perform such an experiment anytime soon because no quantum computer has ever been built. It is likely we will not have the technology needed to build one until some time decades in the future.

The many worlds interpretation of quantum mechanics has become quite popular among cosmologists such as Stephen Hawking, who would like to make use of quantum mechanics to construct theories about the origin of the universe. The other interpretations of quantum mechanics don't really work because they all speak of observations or observers. And, of course, the universe has no outside observers. In the many worlds interpretation of quantum mechanics, this problem disappears. There is nothing in that version that requires observers.

When Hawking speaks about other universes, he uses the technical term "histories." However, he is talking about the same kinds of parallel realities that are discussed by Deutsch. In his book *Black Holes and Baby Universes*, he says, "We happen to live on one particular history that has certain properties and details. But there are very similar intelligent beings who live on histories that differ in who won the war [referring to World War II] and who is Top of the Pops."

Quantum cosmology is an interesting subject, and I will have more to say about it later. For now, I think it would be best to return to the subject of this chapter and look at the implications that the many worlds interpretation of quantum mechanics might have for our understanding of the issues of determinism and free will.

The implications could be significant. Suppose that the kind of experiment that David Deutsch envisions is performed one day, and it confirms that the innumerable parallel universes of which he speaks are indeed real. If that happens, all the questions about determinism and free will that have troubled philosophers and scientists will disappear. It will no longer be possible to ask whether there might be only one possible future. All possible futures will be realized in one universe or another. Similarly, free will would become a non-issue. It will no longer be reasonable to ask whether or not we are compelled to make certain choices, because we will make all possible choices. There will be universes in which you take that job you have been offered, and there will be universes where you do not. There will be universes in which you choose to marry one individual rather than another, and there will be universes in which you never marry at all.

It isn't necessary to try to imagine precisely how events in the quantum world are related to the different paths that out lives will take in the various different universes. If the many worlds interpretation is correct, there would be an unimaginably large number, probably an infinity, of parallel worlds. There would probably also be an infinity of worlds in which more or less exact copies of us lived. Every possible world would be real, and every possibility would be realized in one way or another.

I won't go so far as to say that this must be the case. The whole idea of parallel universes, after all, is based on nothing more than that a certain interpretation of quantum mechanics seems to be possible. There is no empirical evidence to support the idea that these parallel words are real, and there will not be for decades to come, if indeed there ever is. But we can at least speculate that they do exist. After all, physicists like Everett, Deutsch, and Hawking have shown that the idea really isn't all that implausible.

3 | *Is the World There When We're Not Looking?*

"We often discuss his notions on objective reality. I recall that during one walk Einstein suddenly stopped, turned to me and asked whether I really believed that the moon exists only when I look at it." Thus wrote the Dutch-American physicist Abraham Pais, who spent seventeen years at the Princeton Institute for Advanced Studies, where Einstein worked until his death in 1955.

Einstein's question about the moon sounds like the kind of thing that might have been asked of the eighteenth-century British philosopher George Berkeley, who denied the existence of matter, holding that objects became real only when they were perceived. According to Berkeley's philosophy, if it were not for the fact that objects were always perceived by the mind of God, they would have a jerky kind of existence, becoming real only when we looked at them. Why, then, did Einstein ask it of another physicist?

My rhetorical question isn't a difficult one to answer. Einstein asked the question about the moon because most physicists then believed—

and still believe—that the fundamental objects of which our world is composed do become objectively real only when we are looking at them. No, they don't think that tables or chairs or planets or trees possess this strange property. But they do think that such fundamental particles as protons, neutrons, and electrons lack full objective reality. They believe that these particles attain a kind of reality only when they are observed. Physicists think they have exactly the kind of jerky existence that the objects in Berkeley's world would have if there were no God to look at them.

When I say "most physicists," I mean the majority of those who have thought about the meaning of quantum mechanics. I don't include the agnostics, who use quantum mechanics as a tool for making calculations without worrying too much how the picture of reality that the theory gives us should be interpreted. As I have pointed out in chapter 2, it is perfectly possible to do this. Quantum mechanics is an amazingly accurate theory, probably the most successful theory that scientists have ever created. It is the foundation for virtually all of modern physics, and there is no known experimental evidence that contradicts it.

To be sure, anyone who uses quantum mechanics has to deal with such things as particles that spin in both the up and down directions at the same time, or which simultaneously follow more than one path. But it is perfectly possible to do this without wondering what it all "means." If one begins with the right assumptions about some physical phenomenon, and if one performs the calculations correctly, it is virtually guaranteed that quantum mechanics will give the correct answer. There is nothing wrong with this. Some notable physicists have been indifferent to the problem of interpretation. Among them was the British physicist P. A. M. Dirac,* one of the greatest physicists of the twentieth century. Dirac seems to have been interested in quantum mechanics only as a mathematical theory. But this did not prevent him from making numerous important contributions to quantum mechanics during the 1920s and 1930s.

---

*Dirac always used these initials, and not even his closest colleagues knew that they stood for Paul Adrian Maurice.

Nevertheless, ever since quantum mechanics was discovered, there have been physicists who have wondered about the problem of interpreting the quantum mechanical picture of reality. Scientists began pondering the problem during the 1920s, and they are still writing and arguing about it today. Yes, there is an interpretation that has more adherents than the others. But the primary reason for this may be that this was the first interpretation to be developed. As a result, it became more or less the "official" view against which the others had to compete. It was the interpretation that professors taught to their students, if they bothered with interpretations at all. And it was the interpretation that was most often written about.

The Copenhagen Interpretation

This interpretation is called "the Copenhagen interpretation" because it was developed by the Danish physicist Niels Bohr and his colleagues at Bohr's Institute for Theoretical Physics in Copenhagen. Bohr was the first to enunciate it. However, when forming the interpretation, he drew on the ideas of other scientists, notably the German physicist Werner Heisenberg, who had discovered quantum mechanics in 1925.

In 1927 Heisenberg formulated his uncertainty principle, which stated that certain properties of a particle could not be simultaneously determined. One such pair of properties were position and velocity. The more accurately one quantity was measured, the less one could know about the other. And if the position of a particle could be measured exactly, then nothing could be known about its velocity. Similarly, if an experimenter chose to measure the velocity, the position of the particle at the time that the velocity was measured would be completely undetermined.

When physicists state the uncertainty principle, they generally speak of position and momentum rather than position and velocity. I spoke of velocity in the last paragraph to make things a little clearer, taking advantage of the fact that momentum is just mass times velocity. But from now on, I will revert to the ordinary practice. It is always best to stick to standard terminology unless there is some good reason to deviate from it.

When Heisenberg first announced his theoretical finding, it was misinterpreted by some experimental physicists, who took him to mean that experimental technology was not good enough to accurately measure position and momentum at the same time. However, this was not what Heisenberg had said. He had found that it was *mathematically impossible* to determine both quantities simultaneously. It didn't make any difference how good one's measurements were; this barrier would always exist.

"May I Get a Word In?"

The Russian émigré physicist George Gamow once drew a cartoon that depicted Bohr with a colleague who had been gagged and bound to a chair. "Please, please," says Bohr in the cartoon, "may I get a word in?" Of course this was meant as a joke. However, it does illustrate Bohr's method of working on problems in physics. He would generally try to work problems out by discussing them with colleagues. However, the colleagues were often not allowed to say very much; it was always Bohr who did most of the talking.

One of the problems that Bohr and visitors to the institute tackled was the problem of the meaning of quantum mechanics. The uncertainty principle figured prominently in the discussions; nothing like it had ever been encountered in physics before. Some of the implications weren't hard to work out. For example, it was clear that the uncertainty principle was related to indeterminism of quantum mechanics. In Newtonian physics, it was possible—at least in principle—to measure the position and momentum of a body to any desired degree of accuracy. And if one knew the position and momentum of a body, and the forces that were acting on it, it was possible to predict the motion of the body at any time in the future. In quantum mechanics, this couldn't be done; the future motion of a particle like an electron was unpredictable.

Another pair of quantities to which the Heisenberg principle could be applied were energy and time. If one knew the energy state of a particle exactly, then one could say nothing about how long it would remain in that state. This was especially important in the case of atoms, since atoms radiated light when an electron made a *quantum jump* from one energy level to another. If the amount of energy possessed by the electron at each level was accurately known, then it was impossible to say when the jump would take place. It was only possible to speak of the probability that it would happen within a certain period of time. One could know everything there was to know about an atom and be unable to predict when it would emit a ray of light.

All this was clear. However, one question remained unanswered: What was the *meaning* of the fact that quantities such as position and momentum could not be simultaneously determined? Bohr, Heisenberg, and the other physicists with whom Bohr discussed the issue drew the conclusion that a subatomic particle did not have any definite momentum or position except when the particle was observed. The properties of such a particle had no objective reality, except when they were measured.

According to the Copenhagen interpretation, it was not meaningful to speak of the position or momentum of a particle except at times when the particle interacted with an apparatus that had been set up to determine what one quantity or the other was. When one looked at a particle it had certain definite properties, but at other times it faded away—like the Cheshire cat—into a fuzzy cloud or probabilities. According to Bohr, the objects of the quantum world simply did not have the kind of objective reality that we generally ascribe to macroscopic objects. Bohr was uncompromising about this view. "There is no quantum world," he said. "There is only an abstract quantum description."

It was all very well to say that the quantum world didn't exist. But then how would one describe what happened when a measurement was made? Bohr had an answer to this question, too. Before a measurement was made, a quantum system would generally be in a mixture of states that could only be described in terms of probabilities. An electron would have a spin that was both up and down. An electron in an atom would not have any definite energy. It would be in a state that was some kind of ghostly combination of every energy state that it could have. But when the electron or the atom was observed, the combined states would suddenly "collapse," and some specific property would be seen. The electron, for example, would be observed to have spin in either the up or the down direction, not both. And it was not possible to tell before-hand what the observation would reveal.

The idea that a quantum system undergoes a kind of collapse where a bundle of probabilities somehow coalesces into a single reality raised new questions. For example, exactly when does the collapse take place? Does it happen when a quantum object interacts with a piece of appa-ratus? Does it happen when the apparatus registers the result (for example, by displaying a pointer reading)? Does it happen when the experimenter becomes aware of the reading? In the last case, does this imply that consciousness plays a role in causing quantum indefiniteness to become real?

Schrödinger's Long-Lived Cat

Some physicists were deeply skeptical about the Copenhagen interpre-tation. Among them was the Austrian physicist Erwin Schrödinger, the co-discoverer of quantum mechanics. Heisenberg had first discovered quantum mechanics in 1925. In 1926, Schrödinger had discovered a theory of the quantum world that looked somewhat different than Heisenberg's but which was later shown to be mathematically equiva-lent to it. Schrödinger never cared for the approach that was developed at Bohr's institute. He clung to the idea that there was some solid kind of reality underlying quantum phenomena. Subatomic particles were not the seemingly ephemeral things that Bohr claimed they were.

So Schrödinger proposed an imaginary experiment. He hoped that the experiment's apparently absurd implications would show up the flaws in Bohr's reasoning. Schrödinger asked physicists to consider a situation in which a cat was placed in a closed box. Attached to the box were a piece of radioactive material and a Geiger counter. There was a 50 percent chance that the Geiger counter would register a radioactive decay within a certain period of time. If the decay was registered, this would trigger a mechanism that would break a flask of concentrated cyanide gas that had been put in the box with the cat. According to Bohr's interpretation of quantum mechanics, Schrödinger said, the cat must be 50 percent alive and 50 percent dead. After all, the breaking of the flask depended upon a quantum event. It was only when an observation was made, when someone opened the box to see what had happened, that the half-alive, half-dead cat would collapse into one that was either alive or not.

Schrödinger hoped that his seemingly absurd thought experiment would show that there had to be a flaw in Bohr's reasoning. But this isn't what happened. Instead, Schrödinger's cat took on a life of its own, and this imaginary experiment is still being discussed today. Naturally, there have been a number of different interpretations. According to one school of thought, there is no problem: the cat is presumably aware of whether it is still alive or not. This answer is not entirely satisfactory because it raises the problem of what would happen if we put not a cat but an insect or a bacterium into the box. It is hard to imagine that a bacterium "knows" anything, even whether or not it is alive.

The Hungarian-American physicist Eugene Wigner had his own solution to the problem. According to Wigner, the cat's half-dead, half-alive state collapsed when knowledge of this state entered into the consciousness of an observer. Of course, this raises the question of what happens when we put a human being into the box (which we can easily do, since the experiment is imaginary a human is unlikely to come to any harm). If we open up the box and find that he is still alive, he will certainly say that he *never* experienced being in a half-alive, half-dead state. And yet, according to Wigner's interpretation, he was half-alive and half-dead until someone looked.

Schrödinger's cat presents no problems to adherents of the many worlds interpretation of quantum mechanics. According to this interpretation, the cat is alive in some universes and dead in others. If it has been killed, that event has already happened. We simply don't know until we look which kind of universe we are in. Stephen Hawking, who is an adherent of a variant of this interpretation, dismisses the whole problem with a joke. "When I hear of Schrödinger's cat, I reach for my gun," he has said.

Today, some physicists attempt to solve the problem by using the concept of *decoherence*. According to this interpretation, a quantum mechanical collapse takes place when a quantum object interacts with its environment to a sufficient degree. The environment is likely to contain billions of atoms, at the very least, and it is contact with them that makes quantum mechanical probabilities into something concrete. But this theory has its drawbacks, too. It is not so easy to say precisely when the collapse takes place, or how much interaction with the environment is needed before it is likely to happen. And this view doesn't entirely do away with the need for an observer. As far as the observer is concerned, a system is still in a combination of states until a measurement is made.

I am writing this chapter some sixty-six years after Schrödinger invented his thought experiment. Schrödinger's cat may be half-alive and half-dead in a quantum mechanical sense, but as an object of discussion it is still very much alive. It is alive because there is still no universally accepted interpretation of quantum mechanics. As we shall soon see, certain experiments have narrowed the possibilities somewhat. I will be discussing those shortly. But first, it will be necessary to discuss the objections of the Copenhagen interpretation's most celebrated critic.

Einstein's Objections

The most vocal critic of the Copenhagen interpretation of quantum mechanics was Albert Einstein. Einstein initially took issue with the idea that quantum behavior was indeterministic. He frequently expressed his conviction that a new theory would eventually be found that superseded quantum mechanics and restored determinism to the subatomic world. He put forth his view in discussions with Bohr over a decade. From time to time, Einstein would invent a thought experiment by which he hoped to show that nature was really deterministic. Invariably, Bohr would analyze Einstein's experiments, finding some subtle flaw that negated Einstein's argument. This didn't prove that Bohr was right and Einstein was wrong; it only demonstrated that Einstein had not proved his point. Thus the argument went on and on.

A thought experiment is an experiment that can be performed in principle, but which cannot be carried out in reality because measurements of sufficient accuracy cannot be made, or because the necessary experimental technology has not been developed. A thought experiment differs from a real experiment in that it is intended to throw light on some feature of a theory rather than to measure some physical quantity. Thought experiments were frequently used in the 1920s in order to gain a better understanding of quantum mechanics. And in Einstein's hand they became a tool of criticism.

As the years passed, Einstein finally gave up his objections to the indeterminism of quantum mechanics and concentrated instead on an idea that was central to the Copenhagen interpretation: that particles did not possess objectively real properties such as position and momentum. Einstein insisted on a view of reality in which these quantities were real, even though our knowledge of them might be limited.

Theories that posit the reality of such properties have been proposed. They are called *hidden variable theories*, and they embody a point of view that is antithetical to the Copenhagen interpretation. Einstein did not play a role in the development of any such theories, nor did he concern himself with them to any great extent. When he argued with Bohr about quantum mechanics, he was more concerned with the questions about the fundamental nature of the subatomic world than with theory making.

A blow was struck against the hidden variable idea in 1932 when the Hungarian-American mathematician John von Neumann published *The Mathematical Foundations of Quantum Mechanics*, in which he discussed the mathematics of the theory in rigorous detail and gave a proof showing that hidden variable theories were impossible. In this proof von Neumann considered the claim that subatomic particles did indeed have an ordinary kind of objective reality and concluded that this idea was mathematically incompatible with quantum mechanics. Von Neumann's reputation as a mathematician was enormous, and at the time few doubted that the issue had been settled.

But then, in 1952, the American physicist David Bohm succeeded in doing what von Neumann had supposedly demonstrated to be impossible. He constructed a theory in which the electron was regarded as a particle with real properties, and his theory made the same mathematical predictions as quantum mechanics. Bohm's theory didn't attract a great deal of support among physicists; most of them regarded it as too contrived. However, the fact that he was able to construct such a theory showed that something was wrong.

It soon became apparent that von Neumann's "proof" hadn't demonstrated anything at all. It contained an error that should have been glaringly obvious, but which no one had noticed. Von Neumann's reputation as a mathematician had been so great that it had apparently never entered anyone's mind that he could have been guilty of presenting a fallacious proof. But that was precisely what he had done. Hidden variable theories should indeed be possible; questions regarding the reality of subatomic particles hadn't been settled after all.

The EPR Paradox

If Bohm hadn't been convinced by von Neumann's "proof," neither had Einstein. He continued to insist that "the belief in an external world independent of the perceiving subject is the basis of all natural science." In 1935 he wrote a paper with the American physicists Boris Poldosky and Nathan Rosen in which he made yet another attempt to show that there was something wrong with the Copenhagen interpretation.

In this paper, the three authors described a thought experiment highlighting what has become known as the "EPR Paradox." There is nothing paradoxical in the imaginary experiment that Einstein, Poldosky, and Rosen described. They proposed it in order to argue that quantum mechanics was *incomplete*, that there was some underlying objective reality, and that Bohr's description of the behavior of the objects of the subatomic world couldn't be entirely accurate.

The authors of the EPR paper imagined two particles that interacted with each other and then flew apart. Now, the principles of quantum mechanics do not allow us to precisely measure the momentum and position of a single particle. The total momentum of two particles can be precisely known, however, as well as the distance between them. Thus if we know the momentum of one particle, a simple subtraction automatically tells us the momentum of the other, provided that no forces have been acting to change this momentum.

Suppose that, at some instant of time, the position of particle A is measured. Theoretically, this measurement can be made with any required degree of accuracy if we are willing to remain ignorant of the particle's momentum. Next, the momentum of particle B is measured.

This automatically tells us the momentum of particle A. Thus we know both the position and momentum of particle A, in violation of the uncertainty principle. Quantum mechanics cannot account for this, Einstein and his colleagues said. Therefore the theory must be incomplete.

Naturally, Bohr made a reply. But it depended on subtle reasoning, and it most likely didn't seem very convincing to Einstein. Bohr didn't claim to find any error in the EPR argument. He simply maintained that the argument did nothing to overturn quantum mechanics or the Copenhagen interpretation. According to Bohr, what the EPR argument demonstrated was that the idea of separate elements of reality didn't have any meaning in the quantum world, either. Einstein and his colleagues had separated a system composed of two particles into two separate components with well-defined properties. But this was wrong. Such a system had to be considered as an inseparable whole.

After this, communication between Einstein and Bohr more or less came to an end. Each had long been unwilling to accept the other's conception of subatomic reality. And now it was almost as though they had ceased to speak the same language. Einstein believed that he and his colleagues had designed a perfectly clear thought experiment, and now Bohr was saying that breaking a system of two correlated particles into components was an invalid procedure. What was needed was a real experiment to settle the issue. Such an experiment was eventually performed, but it was performed two decades after both Einstein and Bohr were dead.

Bell's Theorem

Einstein died in 1955, and Bohr followed him seven years later in 1962. Neither of them ever had a chance to hear of a remarkable result obtained by the Irish physicist John Bell in 1964. Bell re-analyzed the EPR thought experiment, and found that there might be a way to determine whether Bohr or Einstein had been right.

Like Einstein, Bell had long had doubts about Bohr's interpretation of quantum mechanics. However, he was a theoretical physicist who worked on the design of particle accelerators, not a specialist in quantum theory, so at first he made no attempt to enter the debate. However, after he became aware of Bohm's hidden variable theory, he began to think about the problem of interpretation more deeply. In particular, Bell considered a refinement of the EPR thought experiment that Bohm had proposed. Bohm had simplified the "paradox" by pointing out that only one property of the two particles needed to be considered. If two particles flew away from each other, and their spins were oriented in opposite directions, then a measurement of the spin of one would provide knowledge about the spin of the other, no matter how far away it was. If a measurement showed that one had spin in the up direction, the spin of the other would be down.

Suppose we assign the number +1 to spin in the up direction, and the number −1 to spin down. If two particles had such spins, then the total spin of the system would be zero; this follows from the simple equation $1 - 1 = 0$. So if a particle with zero spin decayed into two spinning particles, the spins of the latter two would *have* to have opposite orientations.

Before I go on to describe what Bell did, it will be necessary to talk a little more about quantum mechanical spin. You may recall that I pointed out in the previous chapter that it is not the same thing as the spin of a macroscopic object. Quantum mechanical spin has the peculiar property that one can measure it in any direction one wants. If the spin is measured in an up-down direction, then each of those orientations will be observed 50 percent of the time. If the spin of the same particle is measured in a right-left direction, then half the time the particle will be found to have a spin pointing left, and the other half of the time will have a spin pointing right. Here I should point out that "right-left," is not an expression used by quantum physicists. I am employing it here because I think it makes it a little easier to see what is going on than it would be if I spoke of y and z coordinates, the terms a physicist would use. But if you feel comfortable picturing particles spinning along y and z axes, you should go ahead and think of them that way.

Naturally, spin doesn't have to be measured in either the up-down or right-left directions. One can choose any orientation one wants. For example, it can be measured in a direction that has an angle of 30 degrees to the vertical. If this is done, there will once again be a

50 percent probability that the particle is spinning in either direction. An experimenter is free to orient his detectors any way he wants.

Quantum mechanical spin has another peculiarity. Suppose that a measurement has been made and a particle has been found to have a spin that is up. If one now measures the spin in a right-left direction, there will again be a 50 percent chance that the spin will be oriented in either of those directions. And if one makes a third measurement, this time of up-down spin, the particle will be found to be spinning up exactly one-half of the time. The second measurement destroys the information that was obtained in the first.

Spin Correlations

Now suppose that two particles with opposite spin are sent flying toward two detectors. If both detectors are oriented the same way, one detector will record a particle with an up spin, and the other a down spin. There will be a 100 percent correlation; the readings taken from the two detectors will always be the opposite of each other. Similarly, if one detector is oriented up-down and the other right-left, then the correlation between their readings will be zero. If one measures an up spin, the other will record a right spin exactly one-half of the time. Finally, if the detectors are oriented at some angle to each other, the correlation will be between zero and 100 percent. For example, if one measures up-down spin and the other is tilted at an angle of 10 degrees to the vertical, they will record opposite readings most, but not all, of the time.

After pondering the problem, Bell saw that an experiment in which the correlation between the spins was greater than zero but less than 100 percent might yield more information than an experiment in which both spins were measured to be up or down. After working out the mathematics, he found that the size of the correlation depended on the assumptions made about the behavior of particles. One set of assumptions would lead to a higher correlation than another. Thus it should be theoretically possible to perform an experiment to determine which assumptions were correct.

Local Reality

One more thing needs to be discussed before we can see exactly what it is that Bell proved, and that is the concept of *local reality*. Local reality isn't an abstruse idea. It is simply a combination of the commonsense view that particles such as electrons are real objects with real properties that continue to exist whether we are looking at them or not and a basic principle of Einstein's special theory of relativity: causal influences cannot travel faster than the speed of light. For example, it takes about eight minutes for light to travel from the sun to the earth. Thus, according to Einstein's theory, something that happens on the sun cannot affect anything that happens on Earth until at least eight minutes have passed.

Bell showed that particles that satisfied the local reality condition would be less highly correlated than those that did not. This implied it might be possible to perform an experiment that would decide between Einstein's and Bohr's views of quantum mechanics. Einstein had maintained that particles satisfied the condition of local reality, while the Copenhagen interpretation depended upon quite a different view. The Copenhagen interpretation does away with the notion that particles are as real as the objects of everyday life. Bell's result, which is known as Bell's Theorem, tells us that it should be possible to perform an appropriate experiment. All one had to do was to set particles flying in opposite directions and measure their spins with detectors that were oriented at an angle to each other, and to position the detectors so that the spins of both particles were measured instantaneously, or nearly instantaneously. If the measurements were performed at exactly the same time, then any influence of one particle on the other would have to propagate at a speed greater than that of light.

Such an experiment would be extremely difficult to perform. When Bell published his findings in 1964, he did not think that it could be done at any time in the foreseeable future. For one thing, no detectors are 100 percent efficient. They sometimes fail to register a particle. This introduces random factors that make correlation difficult to measure. Bell thought of his proposed experiment as yet another thought experiment. However, by the 1970s physicists were already performing experiments along the lines he had suggested.

These experiments used photons, rather than correlated electrons or other correlated particles. Since photons are particles of light, they possess a property called polarization that can easily be detected. Polarized light is nothing more than light that is made up of a lot of polarized photons. Polarization is analogous to spin and pairs of correlated photons can easily be created. For example, calcium atoms can be made to emit correlated photons in opposite directions. The polarization of the photons can then be measured by detectors placed on opposite sides of a laboratory. The detectors are set up at an angle to each other so that they measure polarization in somewhat different directions.

By 1975, six experiments of this type had been performed, and four of them had produced results which indicated that photons did not satisfy the condition of local reality. Since this didn't provide a clear verdict, more accurate experiments were needed. However, the question was finally settled in 1982, when physicist Alain Aspect published the results of experiments performed at the University of Paris's Institute of Theoretical and Applied Optics. Aspect and his colleagues carried out a whole series of experiments over a period of several years, constantly improving the design of their apparatus as they did so. All of the experiments gave the same result. The verdict was against local reality.

It is often said that Aspect's experiments confirmed the predictions of quantum theory. This is true in a sense. But it would probably be better to say that the experiments confirmed any interpretation of quantum mechanics that denied local reality, such as Copenhagen interpretation, *to some extent*. Remember, the idea of local reality has two components: particles have a commonsense kind of reality and influences between them cannot travel faster than the speed of light. Thus the experiments do not rule out the possibility that photons are indeed real particles with real properties that exist whether these properties are measured or not. This could be the case if there were some kind of faster-than-light signaling between photons.

Most physicists do not think this is the case. The idea violates one of the principles of the special theory of relativity, which is as well confirmed as quantum mechanics. Einstein himself had considered the possibility of this kind of faster-than-light signaling and had rejected it, calling it a "spooky action at a distance." There are some very good reasons for thinking that this kind of signaling does not exist. If it did exist, photons would be able to influence one another instantaneously, even if they had flown off to opposite sides of the universe. This seems unlikely. However, the Aspect experiment did not rule the possibility out.

And in fact there is an interpretation of quantum mechanics that does keep the concept of reality while abandoning locality. It is called the *transactional interpretation*, and it was developed by physicist John Cramer of the University of Washington at Seattle. This interpretation is based on the idea that there exist influences between particles that travel both into the future and into the past. The ideas on which it is based are similar to those of the Feynman-Wheeler theory of electromagnetic waves that I described in chapter 1. Recall that Feynman and Wheeler assumed that light and other radiation could indeed travel into the past. The transactional interpretation made the same predictions as other interpretations of quantum mechanics, including the Copenhagen and many worlds interpretations. Thus there is no known experiment that can decide between the Cramer's interpretation and the various others. It appears that, even after the Aspect experiments, we are still free to interpret quantum mechanics in a variety of different ways. In particular, there is still a way to keep the idea of the reality of individual subatomic particles.

If some kind of faster-than-light signaling did exist, it does *not* follow that it would be possible to send messages that traveled faster than light. One could send correlated particles flying off into opposite directions and then make measurements when the particles were light years away. However, no information could be transmitted. If I measure the spin of a particle or the polarization of a photon, I immediately know the spin of a particle that may now be a long distance away. But if someone makes measurements on the distant particle, and the ones that follow it (I am assuming now that whole streams of particles have been created), all he will see is a random sequence of spins up and spins down. For example, he might see something like up, up, down, up, down, down, down, up, and so on. No information could be extracted from that. It is possible for me to know whether his particle is spinning up or down, but I can't make his particle spin in one direction or the other.

Bell's Theorem and Aspect's experiment does establish that Bohr was correct when he said that a pair of correlated particles has to be considered as a system. The properties of one particle depend on measurements made on the other. Such particles are said to be *entangled*. Such particles remain entangled, no matter how far away they may be from one another. It appears that quantum mechanics implies a certain interconnectedness in the universe that is absent from classical physics.

Some authors have attempted to use this interconnectedness to justify the claims of parapsychology. Such arguments should not be taken too seriously. There is no solid evidence for *any* kind of paranormal phenomenon. Parapsychologists have been trying to establish the existence of such things as clairvoyance, telepathy, and precognition since parapsychologist J. B. Rhine began performing experiments at Duke University in the 1930s. No convincing demonstration of the existence of such phenomena has ever been performed. It is not possible to prove that paranormal phenomena *do not* exist. Proofs of nonexistence are generally impossible. It is probably best to maintain a skeptical attitude toward any such claims that are made.

Entanglement is a real phenomenon, but we should not try to read too much into the fact that this phenomenon exists. We know that a measurement made in one place can affect a measurement made in another. This may well have some interesting practical applications. It has been suggested, for example, that entanglement could be used to encrypt messages in such a way that the encryption couldn't be broken. Most likely, entangled particles can be used in ways no one has thought of yet. However, the phenomena of entanglement shouldn't be used to draw far-reaching metaphysical conclusions about the universe, or even more modest conclusions such as the ones that parapsychologists sometimes draw, when these conclusions are not supported by scientific principles.

Well, Is the Moon There When We're Not Looking, or Isn't It?

So far, the Copenhagen interpretation of quantum mechanics has survived all challenges. Bohr was able to find flaws in Einstein's thought experiments. Bohr's interpretation of the EPR paradox turned out to be correct. Presumably someday a quantum computer may detect the existence of the other universes postulated by the advocates of the many worlds interpretation. But that experiment cannot be performed yet. And if it is finally performed, the results may turn out to be ambiguous.

So the Copenhagen interpretation is alive and well, even though there are a number of competing interpretations of quantum mechanics nowadays. I haven't mentioned all of them. The topic of this chapter is the question of whether subatomic particles possess objective reality when not being observed. The chapter isn't intended to be a survey of all the possible ways of looking at quantum mechanics. However, the resilience of Bohr's ideas at least demonstrates that they should be taken seriously.

So what about the moon? Does it exist when no one is looking at it? The moon is a very large object, many many orders of magnitude larger than the particles I have been considering, such as electrons and photons. Yet, like all of the other matter in the universe, the moon is composed of quantum objects: protons, neutrons, and electrons. Consequently the laws of quantum mechanics should apply to it, too.

I don't think you'll be too surprised by the answer I make to these questions. If the Copenhagen interpretation is correct, and the particles that constitute the moon take on reality only when we collectively observe then, then an unobserved moon would fade away into a kind of quantum "fuzziness." Because it is so large an object, this might take billions of years. However, it would inevitably happen.

What Next?

During the 1920s and 1930s, scientists often invented thought experiments in order to better understand the implications of quantum mechanics. Today, thanks to advances in experimental technology, many of these experiments can actually be performed. As a result, quantum mechanics has become the focus of a great deal of experimental research.

Some of these experiments have confirmed that the seemingly strange predictions of the theory are indeed accurate. For example, single neutrons have been made to follow two paths at the same time, and individual beryllium ions (an ion is an atom from which one or more electrons have been removed) have been observed in two places at the same time. It is now possible to trap individual electrons and individual atoms in magnetic fields in order to better study their properties. In interesting experiments performed by American scientists at the State University of New York at Stony Brook and by Dutch scientists at Delft University of Technology in 2000, it was shown that quantum effects are not always confined to the subatomic level; under the right conditions they can be observed in macroscopic objects.

The American and Dutch scientists independently set up experiments in which streams of electrons flowed around superconducting loops. They found they could produce a state in which the electrons were moving both ways—clockwise and counterclockwise—at the same time. This does not mean that half of the electrons were moving one way and half the other way. Each electron was moving in both directions simultaneously. The flows consisted of not just a few electrons, but billions of them. To be sure the loops were not very big. The one at Stony Brook was just large enough to encircle a human hair. However, it was large enough to be seen with the human eye.

You may recall that the Schrödinger's cat thought experiment was intended to be paradoxical because it seemed to imply that a cat—a macroscopic object—could be in two different quantum states at the same time. In 1935, when Schrödinger proposed the experiment, this was thought to be absurd. However, the superconductor experiments conducted in 2000 seem to show that it is not as absurd as was previously believed. A superconducting loop is just as much a macroscopic object as a cat.

In spite of the great amount of experimental activity in the field of quantum mechanics, there are no experiments being planned that could decide between the various different interpretations. There have not even been any suggestions as to how the various interpretations might be tested, with the single exception of an experiment that might be able to test the many worlds version of quantum mechanics. Thus it seems that we will be free to choose whatever interpretations we prefer for some time to come. It is possible to maintain that objects are not there when they are not being observed. And it is possible to maintain the opposite. It is possible to conclude that there is one universe, and it is possible to say that there are probably an infinite number of them. And it is possible to draw a number of other kinds of conclusions as well.

Which interpretation of quantum mechanics is correct? This question hasn't been answered. It is purely a matter of taste. Personally, I suspect that there might be some truth in all of the different interpretations, that they all reveal something about the universe when it is seen from different perspectives. But of course no one really knows.

4 | *Why Is There Something Rather than Nothing?*

Why is there something rather than nothing? Why is there a universe, rather than nothing at all? This hardly sounds like the kind of question science could answer. After all, scientists do not even know why such fundamental particles as protons and electrons exist, or why they have the properties they have. They don't know why objects in the quantum world should behave so differently from macroscopic bodies, and they don't know why light travels at a velocity of 300,000 kilometers (186,000 miles) per second. How could they possibly find a way to understand why there is a universe?

But perhaps matters are not so bad as all that. Quantum mechanics does provide us with some clues. Quantum mechanics tells us that, in our universe, there is no such thing as "nothing." Even completely "empty" space is full of particles such as electrons, photons, and protons that briefly come into existence and then vanish again before they can be observed. These *virtual particles* have extremely short lifetimes. A virtual electron, for example, typically lives for about 10^{-21} seconds

(10^{21} is the numeral 1 followed by twenty-one zeros and 10^{-21} is 1 divided by 10^{21}). This tells us little about the possible reasons for the existence of the universe, but it at least shows us there is something that comes into existence spontaneously. If we want to try to understand the nature of existence scientifically, this might be a good place to start.

The existence of virtual particles is a consequence of Heisenberg's uncertainty principle. Recall that the principle relates not only to position and momentum, but also to energy and time. More specifically it implies that, during short periods of time, the energy of a system must be very uncertain. Thus, during periods of time that are short enough, there should be enough energy available to create particles out of nothing. This is a consequence of Einstein's famous equation $E = mc^2$, which says that matter and energy are equivalent; under the right circumstances, one can be converted into the other.

Particles such as electrons cannot be created individually. The reason for this is that the electron has a charge, and certain other properties, such as spin. Mass can be created out of nothing, but electric charge cannot be. Neither can spin. If an electron is created, another particle with opposite spin must be created, too. When virtual particles come into existence, they are always created as pairs. An electron and a positron may come into existence, for example. The electron has a negative charge; the positron can be described as a positively charged electron.

The positron is said to be the *antiparticle* of the electron. Particle-antiparticle pairs can be created out of energy. And when a particle and its antiparticle meet they annihilate each other and a small burst of energy appears in their place. Annihilation is the fate experienced by virtual particles. When this annihilation occurs, the energy that the electron and positron "borrowed" from empty space is paid back. The process by which a virtual particle pair is created and then destroyed is given the name *quantum fluctuation*.

Quantum fluctuations happen everywhere: in interstellar and intergalactic space, in the computer on which I am writing this book, in the desk on which it rests, inside the sun, inside your head, and inside the banana you are thinking of eating. Quantum fluctuations are a ubiquitous feature of the universe in which we live. It would not be possible to create conditions under which they did not take place, no matter how sophisticated the technology that was available.

Electrons and positrons are not the only particles that appear in quantum fluctuations. For example, photons have zero mass, but they do have energy. Thus pairs of virtual photons are also constantly appearing and disappearing. The photon has the property of being its own antiparticle, so pairs of photons constantly pop into existence for brief times, and then disappear. Similarly, empty space contains large numbers of virtual neutrinos and antineutrinos. The neutrino weighs considerably less than the electron, hence it is created in much greater numbers. On the other hand, virtual particles heavier than the electron, such as neutrons and protons, are relatively rare. Because they have larger masses, more energy is required to create them, and this energy must be "paid back" in a shorter period of time.

Don't be misled by the name "virtual." Virtual particles may be short-lived, but they are very real. Although they cannot be observed, their existence has been confirmed experimentally. One such experiment is one that was suggested by the Dutch physicist Hendrik Casimir in the 1940s. Casimir pointed out that if two metal plates were placed very close to each other, then there would be electromagnetic fields associated with the virtual photons in the space between the plates, and these fields would create a force of attraction between the plates. The force is small and the experiment is thus difficult to perform. However, in 1993, physicist Ed Hinds and his colleagues at Yale University did succeed in measuring the force. They found that the predictions of quantum mechanics were confirmed.

Another experimental confirmation of the existence of virtual particles is somewhat more indirect, but even more convincing because it confirms the predictions of quantum mechanics to such a high degree of accuracy. Every electron is a tiny magnet, and an electron's magnetism can be measured to a very high accuracy. If virtual particles are constantly popping into and out of existence in the vicinity of an electron, then the magnetism of the electron will seem to be somewhat different than it would be if virtual particles did not exist. The magnetism of an electron surrounded by virtual particles has been calculated, and the calculation agrees with experiments to an accuracy of better than one part in 10 billion. Hence we can conclude that the existence of quantum fluctuations is well confirmed.

The Big Bang Theory of the Origin of the Universe

It might seem that the existence of quantum fluctuations is not all that relevant to our understanding of the universe. After all, quantum fluctuations take place on the submicroscopic level, and the universe is very, very large. Nowadays astronomers commonly observe galaxies that are many millions of light years away, and the visible universe is presumably only a small part of a larger whole. However, this is not the case. At one time, the universe was very small and quantum processes had significant effects. A knowledge of these effects is vital if one is to attempt to understand the early stages in the evolution of the universe, or to speculate about its origin.

Before I discuss the very early universe, it will be necessary to explain the standard theory of the evolution of the universe, the big bang theory. This theory has been discussed numerous times in numerous different books. If it is reasonably familiar to you, there is no reason why you should not skip over this section. I will only be discussing the basics in the next few pages. Discussion of recent discoveries and speculation will come later.

Hubble's discovery of the expansion of the universe in 1929 ushered in a new era in cosmology. Once it was realized that the universe was not static and unchanging, it became possible to speculate about its history. In particular, scientists began to realize that if it was expanding today, then there must have been a time when all of the matter of the universe was compressed into a relatively small space. After all, the only force that seemed to be acting on the universe as a whole was gravity, which simply acted to slow down the expansion. This implied that the universe must have been expanding from the very beginning.

In 1933, the Belgian astronomer and cosmologist Georges Lemaître proposed the first theory of the origin of the universe. He hypothesized that all of the matter in the universe was originally compacted into a cold "primeval atom," which then disintegrated by a process resembling nuclear fission, sending matter flying off in all directions. The current expansion of the universe, he speculated, was the result of this initial explosion.

Today we know that Lemaître's idea was correct in principle, and wrong in most of its details. The universe never existed as the kind of solid ball that Lemaître conceived. And it was not initially cold, as he thought. On the contrary, it was very hot. For example, at a time of one second after the beginning of the big bang, the universe had a temperature of about 10 billion degrees Celsius. It was very hot then, and it has been cooling off ever since. But even though many of Lemaître's conclusions were wrong, he has to be considered the father of modern cosmological theory.

An improved version of the big bang theory was developed by George Gamow and his colleagues in the 1940s. According to Gamow's hypothesis, the universe was originally made up of neutrons packed together in a fireball that had a temperature of well over 1 billion degrees C. Some of the neutrons decayed into protons and electrons, and the particles in the fireball combined to produce the elements that are observed in the universe today. This was much closer to the truth, and what is called the big bang theory today is essentially a modification of Gamow's hypothesis.

It is no longer believed that the universe was once made up of only neutrons. Scientists now know that many different kinds of particles were present in the big bang fireball, and that protons and neutrons must have been present in roughly equal quantities. However, Gamow's idea that the universe we observe today had its origin in a hot "soup" made up of subatomic particles interacting with one another has been shown to be correct.

Today the big bang theory is supported by a number of different kinds of evidence. For one thing, the light from the primeval fireball can still be seen in the form of microwaves (short wavelength radio waves) that fall on the earth from every direction of space. These microwaves are called the *cosmic microwave radiation*. They were emitted about 400,000 years after the big bang, when the universe became cool enough that electrons could combine with nuclei to form neutral atoms. The big bang took place about 13 or 14 billion years ago, and in those billions of years what was originally visible light has been transformed into radiation that is much weaker. In one sense what we see today is like a dying ember of a once-very-hot fire.

It isn't difficult to understand why we should see what was once light as microwaves. The universe has been expanding for billions of years. As space expanded, the wavelengths of this radiation were "stretched" by the same amount; the distance between successive wave crests increased. The wavelengths of microwaves are many orders of magnitude longer than those of light. What was originally light became the low-energy radiation that scientists observe today.

Next, there is considerable evidence that 13 or 14 billion years ago the universe was composed of a mixture of hydrogen and helium, with small traces of some other light elements. It has been determined that the universe was about 23 to 24 percent helium by weight. Scientists can determine the chemical composition of stars, clouds of gas, and other objects in the universe by studying the light and other radiation they emit. No star, galaxy, or cloud of gas has ever been observed where the helium content was less than this quantity. This helium must have been present at the beginning because such a large quantity of helium could not have been manufactured in stars. There has not been enough time. For example, the surface layers of our sun, which has been shining for about 5 billion years, contain 27 percent helium. The extra 3 or 4 percent is just what one would expect to see in a star of the sun's size and age.

If the helium that astronomers observe was not made in stars, there is only one place that it could have been produced: in the big bang fireball. The nuclear reactions that took place then must also have been responsible for the existence of certain trace substances that cannot be manufactured in stars at all. One of these is deuterium, or heavy hydrogen. A hydrogen nucleus is simply a proton, but a deuterium nucleus is made up of a proton and a neutron. These two particles are weakly bound together. Not only can deuterium not be made in stars, the nuclear collisions that take place within a star's interior would quickly break it apart. The concentration of deuterium in the universe is 15 parts per million by weight. This deuterium could only have been manufactured in the big bang fireball.

Finally, if there was a big bang, the universe must be constantly cooling, in a manner similar to the cooling of a gas when it expands. If you have ever caused gas to be released from an aerosol can and have felt the can become cold, you have observed this effect. Today, the temperature of the universe is about 2.7 K. Here K stands for kelvins and denotes degrees Celsius above absolute zero. Absolute zero, the temperature at which molecular motion ceases, is the lowest possible temperature, and is about −273 C. You may be a bit surprised at this low figure. After all, the universe contains numerous stars, and stars are very hot. However, stars occupy only a tiny fraction of the total volume of the universe. If all of the matter in the universe were "smeared out" uniformly, its density would be less than one atom per cubic meter. Thus stars have little effect on the overall average temperature.

If the temperature is 2.7 K today, we should expect to find that it was greater in the past. And indeed this is just what is observed. When astronomers look far out into space, they are also looking far back in time. A galaxy that is 5 billion light years away is seen by light that it emitted 5 billion years ago. This follows from the definition of "light year," the distance that a ray of light travels in one year. Thus it is possible to directly observe the universe at past epochs.

In 2000, a team of Indian, French, and German astronomers succeeded in analyzing the light absorbed by a cloud of gas and dust when the universe was only one-sixth its present age. When an object absorbs light, the light's energy is taken up by individual atoms, which go into energy states that are higher than normal. The multinational team of astronomers found more atoms in higher energy states than are seen in similar clouds near our solar system. They concluded that there was only one possible explanation for this. At the time the light was being absorbed, the radiation temperature of the part of the universe surrounding the cloud had to have been between 6 and 14 K. This was consistent with the big bang theory, which says that the universe had a temperature of 9 K at that time, more than three times the 2.7 K that is observed today.

The Inflationary Universe

Gravity acts as a retarding force on the expansion of the universe. If the density of matter in the universe is less than a certain amount, called the *critical density*, the universe will expand forever. If the density is more than that amount, then the expansion of the universe will eventually cease, and a phase of contraction will ensue. And if the density of the universe is exactly the critical figure, the expansion will gradually slow down, but it will never quite come to a halt.

Here I am ignoring the recent discovery that the expansion of the universe is accelerating. Although this implies that the expansion will always continue, it does not much affect the points I will be making about critical density. The cosmic force that causes the acceleration of the expansion is significant today, when the universe is many billions of years old. However, it did not significantly affect the expansion of the universe at earlier epochs, when the universe was more dense and the expansion more rapid, and the retarding force of gravity was stronger. It was stronger because the bodies in the universe were closer together, making the gravitational attraction between them stronger.

This critical density is approximately equal to five atoms per cubic meter. The quantity of visible matter in the universe, that contained in stars, galaxies, and objects such as gas clouds has an average density of only 0.2 atoms per cubic meter, about 4 percent of the critical figure. Thus one might think that it is pretty obvious that our universe would expand forever, even if there were no dark energy causing the expansion to accelerate.

Indeed it would be possible to draw this conclusion, if visible matter were all that the universe contained. However, during the last several decades, scientists have learned that there is a lot of matter in the universe we can't see. This matter emits no light, or radio waves, or any other kind of radiation and consequently cannot be detected. We know that this *dark matter* exists because, even though astronomers cannot see it, they can observe its effects. The gravitational forces exerted by this matter influences the motion of objects they can see.

Some of the dark matter undoubtedly exists in the form of *brown dwarfs*. Brown dwarfs are objects that are massive, but not massive enough to ignite as stars. When a star is formed, gravitational forces cause it to contract. This contraction causes the pressure to rise in the protostar's interior, and temperatures arise. In a brown dwarf, the temperatures do not become high enough to ignite the nuclear reactions that take place in stellar interiors.

Since brown dwarfs cannot exist in great enough numbers to account for all of the dark matter observed, scientists believe that much of it consists of exotic particles left over from the big bang. No one is sure precisely what these particles are, although a number of different possibilities have been suggested. Most likely, the dark matter is made up of particles that have not yet been seen in laboratories on Earth.

Dark matter exists in sufficient quantities that the total matter density of the universe must be very close to the critical value. You may not find this fact very surprising, but astronomers and cosmologists were surprised by it once they realized its significance. As the universe expands, deviation from the critical density is magnified. The closeness of the universe to critical density today implies that at a time of one second after the beginning of the big bang, the density of the universe must have equaled the critical density to an accuracy of one part in 10^{15}.

The standard big bang theory does nothing to explain why the density of the universe is so close to the critical value. It hardly seems credible to imagine that this came about through some cosmic coincidence. Some sort of explanation is required if we are to understand why the universe is the way it is.

The problem was solved in 1980, when the American physicist Alan Guth proposed his inflationary universe theory. Guth found that certain theories of subatomic particles, called *grand unified theories*, or GUTs, seemed to imply that when the universe was a tiny fraction of a second old large forces would have existed that would have caused the universe to expand at a rapid, "inflationary" rate. According to Guth's theory, the expansion began when the universe was about 10^{-35} seconds old, and continued for a time of 10^{-32} seconds.* In this brief period of time, the universe increased in size by a factor of 10^{50} or more. When the forces

---

*10^{32} is the number represented by the numeral 1 followed by 32 zeros; 10^{-32} is 1 divided by 10^{32}. It is an extremely small positive number.

driving the inflationary expansion dissipated, the expansion slowed down, and continued at a more leisurely rate, creating the relatively slowly expanding universe that we see around us.

At first there was no direct observational evidence in support of the inflationary universe theory. Nevertheless, the theory became widely accepted for a number of different reasons. First, the theory explained a number of things about the universe that no other theory was capable of explaining. For example, calculations showed that inflation would have fine-tuned the rate of expansion of the universe to a value that was precisely what was observed. Second, further theoretical research showed that an inflationary expansion would begin under a wide variety of different conditions. Inflation does not seem to depend upon the particular theories that Guth made use of in the original version of his theory. The idea of an inflationary universe has become so popular, in fact, that there are now more than fifty different versions of the theory. They all say that an inflationary expansion should have taken place, but differ in their detailed descriptions of the processes that were taking place.

In 2001 the first direct observational confirmation of the theory was obtained. A large international team of scientists observed the microwave background with a radio telescope, called Boomerang, which had been launched by balloon above the South Pole. They observed small fluctuations in the background that were of precisely the size that the inflationary universe theory predicted. According to the theory, an inflationary expansion would have inflated quantum fluctuations in the very early universe to macroscopic size. This would create temperature fluctuations in the universe at the time the radiation was emitted, 400,000 years after the big bang. The Boomerang group's studies of the cosmic microwave background indicated that these temperature fluctuations had indeed existed, and that they were of the size the theory predicted.

Is the Universe a Quantum Fluctuation?

If the inflationary universe theory is correct, then the universe probably contained only about 25 grams, or approximately one ounce of matter at the time the inflationary expansion began. Most of the matter that exists today—the matter that became the stars and galaxies we see today—was created during the period of inflation. It would not be inaccurate to picture both matter and energy as rushing in to fill the rapidly expanding space.

This raises the possibility that the universe may have originally contained no matter at all, and consequently that it may have originated from nothing. Indeed, precisely this proposal was put forward by the American physicist Thomas Tryon in 1973, seven years before the inflationary universe theory was discovered. Tryon suggested that the universe might have begun as a quantum fluctuation in empty space, that the entire universe, and all the matter and energy it contained, materialized from the vacuum. Tryon based his idea on the fact that the total mass-energy content of the universe appears to be zero. The universe contains a great deal of matter and a great deal of energy. But most of the energy is gravitational energy, which is negative. Furthermore, there are reasons for thinking that the positive matter and the negative energy exactly balance each other. If this is the case, it is indeed conceivable that our universe could have come from nothing. As Tryon has said, our universe may be "simply one of those things that happens from time to time."

It is not too difficult to see why gravitational energy has to be negative. Suppose that a space vehicle is sent from Earth to some distant point in space. Energy is required to propel the vehicle upward from Earth. Thus, when it has reached a height of, say, ten miles, it must have more gravitational energy than it had when it was on the surface of Earth. Next, consider the situation where the space vehicle is so far away from Earth that it no longer feels Earth's attraction. Since there is, in effect, no gravity, its energy must be zero. But if it had *less* energy than this when it was on Earth's surface, its energy must then have been negative. Similar arguments apply to any pair of objects that are close enough to one another to experience measurable gravitational attraction. The universe contains an enormously large number of objects that attract one another gravitationally. For example, there are about 100 billion stars in a spiral galaxy like our Milky Way, and there are at least 100 billion spiral galaxies. But, as we saw previously, visible matter such as stars makes up only a small portion of the total matter in the universe.

When matter is created from energy, it is created in the form of particle-antiparticle pairs (if the energy is really there, then real particles—not virtual particles—are produced). In our universe, particles greatly outnumber antiparticles. Antimatter—matter made of antiparticles—has never been observed. However, this is no impediment to Tryon's hypothesis. It is believed that in the early universe—at a time when it was much less than one second old—particles and antiparticles existed in equal numbers. These were not all the kinds of particles that physicists see today. The high energies that were present in the early universe would have made it possible for a great variety of exotic particles to be created. If some of these decayed into other particles slightly more often than they decayed into antiparticles that would explain the character of the present-day universe.

The asymmetry would not have to be great. If a billion and one particles were created for every billion antiparticles, then the billion antiparticles and a billion of the particles could mutually annihilate, and all that would remain would be particles of ordinary matter. These would eventually form all the galaxies, stars, planets, and other matter that exist in the universe today.

In 1967 the Russian physicist Andrei Sakharov conceived of a theoretical mechanism that could produce a tiny excess of matter over antimatter. His theory required, among other things, that there had to be particles that exhibited time-asymmetric behavior. If a particle had this property, there could be slight differences between the decay of a particle and its antiparticle, and this could lead to the existence of more particles than antiparticles. The recent observations of the behavior of the B meson indicate that some particles do indeed behave in this manner, lending support to Sakharov's theory. Thus there is good reason to believe that the universe need not have contained more matter than antimatter originally.

In some respects, Tryon's hypothesis of creation out of nothing wasn't implausible, but it wasn't exactly a theory of the creation of everything. Tryon proposed that the universe had arisen from empty space, yet his idea didn't explain the existence of this space. As we saw at the beginning of this chapter, empty space is not "nothing." It is full of virtual particles. Furthermore, these virtual particles have energy. It appears that if we want to accept Tryon's idea, then we have to conclude that our universe came from another, much larger universe that was devoid of matter but which nevertheless contained energy. Tryon's theory said nothing about the origin of this prior universe.

Nevertheless Tryon's hypothesis caught the interest of other scientists, who proposed variations on the idea. In 1978, four Belgian physicists published a paper in which they suggested that the universe might have begun as a single particle-antiparticle pair. And in 1981, physicists Heinz Pagels and David Atkatz of Rockefeller University in New York, suggested that the universe might have begun with a sudden change in the dimensionality of space. According to their hypothesis, the universe initially contained no matter and had a large number of spatial dimensions. The big bang could have taken place, they suggested, when the universe suddenly "crystallized" into its present form.

All of these hypotheses depended on the existence of an unexplained "something" that preceded the universe. But they did not explain why this something existed, or why it had the properties it had. But then, in 1982, Tufts University physicist Alexander Vilenkin suggested that the universe might have been arisen from "literally *nothing*." According to Vilenkin, a quantum fluctuation might have created not only matter but space and time as well. He showed that the idea that the universe sprang into existence out of absolute nothingness was consistent with quantum mechanics and with general relativity. If Vilenkin's idea is correct, then the universe might have begun when a tiny bubble of spacetime spontaneously appeared and then underwent an inflationary expansion. And of course, if this happened once, it could happen numerous times. There could be an innumerable number of universes, and additional universes might be created all the time.

Stephen Hawking has also suggested a mechanism by which the universe could have come into existence. In 1983, he and the American physicist James Hartle published a paper in which they proposed an alternative idea. According to Hawking and Hartle, it was possible to construct a quantum-mechanical description of the universe in which at

very early times, the time dimension resembled a dimension of space.*
In effect, there were originally four spatial dimensions, and no time at
all in the sense that we know the latter. At later "times," time would
behave in a normal manner.

Hawking's and Hartle's hypothesis, called the *no-boundary proposal*,
differs from Vilenkin's in one important respect. According to their
hypothesis, there was no moment of creation. The universe does not
come into existence out of nothing, because "nothing" never enters the
picture. Such a universe, Hawking has said, "would neither be created
nor destroyed. It would just BE."

All of these proposals are highly speculative, and all of them have an
unavoidable drawback: they depend on quantum mechanics and the
general theory of relativity! Although both are extremely well-confirmed
theories, they are inconsistent with each other. In particular, general relativity breaks down in regions smaller than about 10^{-33} centimeters,
where quantum effects may alter the nature of space and time themselves. Normally this does not cause any problems. However, at a very
early time, prior to about 10^{-43} seconds after the beginning, the universe
was smaller than this. Furthermore, at this time, which is called the
Planck era after the German physicist Max Planck, the strength of the
gravitational forces exerted by individual particles would have been
comparable to electromagnetic and nuclear forces. And there is no
known theory capable of describing such a situation.

A theory that combined general relativity and quantum mechanics
would presumably be capable of describing phenomena that took place
under such circumstances. However, the incompatibility of the two theories has defeated all attempts to create a theory of quantum gravity.
Thus hypotheses about the origin of the universe have a shaky theoretical foundation.

This has not put a damper on speculation, however. Although none
of these proposals can be empirically tested, they do show it is plausible
that our universe could have been created spontaneously, possibly by a

---

*In his book *A Brief History of Time*, Hawking spoke of "imaginary time" when he described
this hypothesis. This may have confused lay readers who didn't realize that the word "imaginary" was being used in an abstract mathematical sense, and not in the way that we use the
word in everyday life.

quantum fluctuation of one kind or another. In fact, an entire new field of physics called *quantum cosmology* has been created by scientists such as Hawking, in order to deal with questions concerning the origin and evolution of the universe.

Quantum cosmology is based on a simple idea: if quantum mechanics is a correct theory, then it ought to be possible to apply it to the entire universe. The Hawking-Hartle no-boundary proposal is a good example of the kind of idea that quantum cosmologists have put forward. Although is it is possible to go back in time and observe what was happening when the universe was very young, a theory in quantum cosmology can be checked to a certain extent by trying to see if it would produce a universe similar to that which we observe today. Thus if a hypothesis turned out to imply that the universe was devoid of matter, or that it had other properties unlike those that we observe, it would be easy to conclude that there had to be something wrong with the initial assumptions.

One generally thinks of physicists as people who try to find out what the world and the universe are like. However, speculation, even very wild speculation, plays a role, too. Scientists speculate in order to understand what might be possible. Finding out what possibilities are at least reasonable, and which are unreasonable, is an essential part of the advance of science.

Something out of Something: Creating Universes in the Laboratory

During the mid-1980s Alan Guth, in collaboration with his students and colleagues, began to investigate the question of whether universes could be created in the laboratory. No, he was not seriously proposing that this might be possible with present-day technology. He admits that it is "beyond the range of any *conceivable* [his italics] technology." The intent of his theoretical research was to try to see under what conditions a universe might come into existence in some previously existing universe.

Guth realized that trying to create a universe at the instant of its big bang wasn't workable. According to the general theory of relativity, the mass density, pressure, and temperature were all infinite at that time. This isn't a conclusion we should take literally, by the way. It's simply an indication that general relativity breaks down at very early times. As I have previously pointed out, the theory simply cannot describe what was going on during the Planck era.

Guth realized it wouldn't do to begin with a universe that resembled ours at a time of one second, either. You may recall that when our universe was one second old, it had a temperature of about 10 billion C. All of the subatomic particles it contained possessed a great amount of energy, and it is possible to calculate that the mass/energy content of a universe one second old would be about equal to the energy in 10 billion universes like ours. Even if one could use the energy in 10 billion universes to make one new one, it would not be a very efficient process.

So Guth and his various collaborators began by considering the creation of a universe that was just about to undergo an inflationary expansion. Such a universe would measure 10^{-26} centimeters across, and would contain about an ounce (25 grams) of matter. A distance of 10^{-26} centimeters is a very small distance; it is about 10 trillion times smaller than the diameter of an atomic nucleus. A universe of this size that contained 25 grams of matter would have to have a density of 10^{80} grams per cubic centimeter. This is an enormous quantity (remember that 10^{80} is "1" followed by eighty zeros). By comparison, the density of water is 1 gram per cubic centimeter.

However, no one knows what technologies will be available in the far distant future, so one might as well go ahead and assume that an object with this kind of density could be created. It is then possible to try to imagine what would happen next.

I won't give all the details of Guth's idea. They can be found in his book, *The Inflationary Universe* (Addison-Wesley, 1997). It should be sufficient to say that Guth and his collaborators concluded that it was likely that the newly formed universe would "bud off" from our universe and quickly disappear from sight. At first, the child universe would be connected to our universe by a microscopic wormhole. The wormhole would then pinch off, and all connections between the two universes would be severed. According to Guth's calculations, the new universe would disappear in about 10^{-23} seconds, releasing a quantity of energy equivalent to a 500-kiloton nuclear explosion. Obviously, creating a universe is not something that you would want to do in your basement.

A wormhole is a highly distorted region of spacetime, a kind of neck that can connect two universes or two widely separated regions of space. Wormholes have never been observed, though their existence is allowed by the general theory of relativity. It is unlikely we will ever encounter wormholes like those in some of the *Star Trek* television series. If wormholes exist, they are almost certainly submicroscopic and much too small to be seen.

Guth does not claim that creating a universe is theoretically possible. In fact, he cites reasons why it might turn out to be theoretically *impossible*. And the fact that no theory of quantum gravity exists introduces a number of uncertainties. However, Guth's careful study of the ways in which universe creation might be carried out have improved scientific understanding of the processes that would be taking place if a universe were created in an existing universe.

So Why *Is* There Something Rather than Nothing?

The answer to this question is that we really can't be sure. It has been shown that the idea that the universe may have popped into existence out of nothing has some plausibility. But there is no evidence that this is indeed what happened. Scientists who ponder the problem are hampered by the lack of a theory of quantum gravity that could describe what might have been happening at the instant of creation and shortly thereafter. Until such a theory is found, they will be forced to rely on guesses to a certain extent.

Note that if Hawking's idea is correct and the universe "just IS," this question may not even have an answer. Hawking and Hartle's no-boundary universe has no beginning in time. According to their proposal, time as we know it did not originally exist. A Hawking-Hartle universe is one that cannot be created because it is a universe that has no beginning. Nor could it have an end. If the universe eventually entered a phase of contraction (this seems unlikely but cannot be ruled out with 100 percent confidence) and collapsed in a big crunch, the same no-boundary conditions would be created. Such a universe would have neither a beginning nor an end. It would evolve from a state where there were only four spacelike dimensions to an identical state. In such a universe, time would be something that existed only temporarily.

The scientists who speculate about the origin of the universe are also hampered by the fact that we have direct observational evidence about the state of the universe only down to a time of one second. Deuterium and other trace elements were created when the universe was one second to three minutes old. Measurement of their abundance in the universe provides scientists with information about the processes that were taking place then. It is believed that the temperature, pressure, and matter density of the universe at this time are known with reasonable accuracy. At times before one second, scientists must rely on theory: quantum mechanics and the general theory of relativity. It is thought that general relativity should be accurate down to a time of 10^{-43} seconds. However, this assumption has never been tested. We cannot reproduce the conditions—the high temperature, pressure, and matter density—that existed at that time. And, of course, there is no theory at all that is capable of dealing with times before 10^{-43} seconds.

But this doesn't imply that the title question of this section cannot be answered. Many physicists believe that a theory that describes both quantum phenomena and gravity can be developed, and intense theoretical research along these lines is currently being conducted. There is a class of theories called superstring theories that do promise to provide us of an understanding of the phenomena that would take place under extreme conditions, such as those that existed at very early times.

I will have more to say about superstring theories later. For now, a brief explanation should suffice. Superstring theories picture particles as tiny vibrating loops in a multidimensional space. Standard superstring theory posits ten dimensions: nine dimensions of space and one of time. According to these theories the extra spatial dimensions are *compacted*, or rolled up, to dimensions many orders of magnitude smaller than an atomic nucleus. If you could travel all the way around the universe along one of these dimensions, your journey would be infinitesimally short.

In order to picture a compacted dimension, do the following: Imagine that a sheet of paper is rolled up into a tube, and that the tube is then tightened, causing its diameter to shrink. The tube becomes thinner and thinner, and if this process could be carried on indefinitely, the tube would eventually become almost indistinguishable from a one-dimensional line. One of its two dimensions could then be said to be compacted.

A great deal of theoretical effort has been put into superstring theories, and none of these theories have yet yielded a single experimentally testable prediction. These theories are horrendously complicated mathematically, and progress has been slow. At present, physicists do not even know how the theories can be made to predict the particles that are observed, such as electrons, quarks, neutrinos, and so on. These particles are presumably nothing but different vibrational states of the superstring loops. However, no one knows how the loops can be made to vibrate in the appropriate ways.

In spite of all these difficulties, many physicists believe that a superstring theory will sooner or later fill in the gaps in present-day theoretical knowledge. The appeal of these theories stems partly from the idea that they automatically incorporate gravity into their description of nature. If the efforts of the superstring theorists are successful, the problem of the incompatibility of general relativity and quantum mechanics would disappear.

Thus a successful superstring theory might well tell us what was happening during the first 10^{-43} seconds—and perhaps at the moment of creation itself.

5 | *Could the World Be Otherwise?*

Obviously, there are ways that Earth *could* be. If a collision of Earth with an asteroid had not caused the extinction of the dinosaurs 70 million years ago, we would not exist. At the time that the dinosaurs died, the largest mammals were only about the size of a ferret. Mammals lived in the ecological nooks and crannies of a world dominated by dinosaurs. Most likely this would still be the case if they had not been given the opportunity to expand into the ecological niches that the dinosaurs vacated.

There are many occasions on which evolution could have followed a different course. The extinction that took place 65 million years ago is only one of many that Earth has experienced. There have been numerous others, in which as many as 95 percent of the species inhabiting this planet have perished. Mass extinctions are only one way in which the element of chance enters into evolution. It is easy to imagine many ways in which evolution could have happened differently.

Earth and the solar system could have been formed differently. For example, if Earth had not experienced a collision with an object about the size of Mars billions of years ago, we would not have a large moon. Only a collision of that magnitude could have knocked a moon-sized chunk of matter off Earth. And of course, Earth could have been a little closer to, or a little farther from, the sun. The sun was formed from interstellar gas, mostly hydrogen and helium. If there had been a little more or a little less of the gases present, the sun might be smaller or larger than it is.

Chance seems to play a role almost everywhere we look. There are a lot of ways in which human civilization could have followed a different course. Our world would be quite different if migration and colonization patterns had been other than what they were. History would have been altered if certain rulers had lived longer, or had died sooner, or if certain battles had been won rather than lost. It would be different if certain innovations had not happened when they did. For example, the invention of the horse collar transformed European civilization by making it possible to use horses as draft animals. History would have been different if, near the end of the eleventh century, Pope Urban II had not conceived of a crusade to conquer Palestine. And he most likely would not have gotten this idea if the Byzantine Empire had not appealed to Rome for aid against the Turks, who had been harassing their borders.

Examples like these can be multiplied endlessly and raises the question of whether there is anything at all that is inevitable. If the character of the sun and Earth, the course of evolution and the evolution of intelligence, and human history all depend on chance events, is there anything that could not have been otherwise?

The Laws of Nature

Are the laws of nature also a product of chance? There is no obvious reason why they should not be. It is easy to imagine universes in which physical laws are not the same as the laws in our universe. Or could there be some fundamental reason why they are as they are? This is the question that Einstein had in mind when he asked whether "God could have made the universe in a different way." Einstein was speaking metaphorically when he said this. What he was asking was whether the laws of physics could have been different, or whether the requirements of logic and internal consistency dictated their form.

Most likely Einstein asked this question because he felt there *was* a certain inevitability about the laws of physics. Einstein was led to many of his discoveries because his keen intuition told him that the universe must be a certain way. This is reflected in some of the other remarks he made. For example, the first confirmation of Einstein's general theory of relativity was obtained in 1919 when British astronomers observed the sun during an eclipse and discovered that the path of starlight that grazed the sun's surface was indeed bent, as the theory had predicted. Apparently Einstein appeared unmoved when he heard the news. When one of his students, Ilse Rosenthal-Schneider, asked him why he wasn't as excited as she was, Einstein replied, "But I knew that the theory is correct." When she asked him how he would have reacted if the theory had not been confirmed, he answered, "Then I would have been sorry for the dear Lord—the theory is correct."

There does seem to be a certain inevitability about some of Einstein's ideas. Both his special and general theories of relativity are based on simple ideas. The special theory depends on the postulate that the speed of light, measured by any observer, is always the same. The general theory, Einstein's theory of gravity, is based on the assumption that the effects of acceleration and gravity are indistinguishable. For example, if an astronaut is in a space vehicle that is accelerating at 1 G, everything will seem to be exactly the same as it would be if his vehicle were standing motionless on Earth. If he had no instruments that told him whether or not he was moving, it would not be possible to distinguish a 1 G acceleration from a 1 G force exerted by Earth.

Einstein's greatness stemmed from his ability to take ideas like these and work out their implications in detail. The ideas from which he began do seem to be almost self-evident, and the theories he derived from them exhibit a great deal of logical consistency. So perhaps it was inevitable that Einstein should have felt the way he did about the inevitability of the laws of physics.

On the other hand, most physics is not like Einstein's. Numerous physical phenomena cannot be explained in this manner. It is often necessary to try out assumptions that are anything but obvious, and there exist complex phenomena that can only be explained in complex ways. As a result, most theories in physics do not have the same aura of inevitability that Einstein's had. And some of them are not quite as clear and logical as his are. For example, *quantum electrodynamics*, or *QED*, the theory that deals with the interaction of light and matter, might not even be mathematically consistent. The predictions of the theory are extraordinarily accurate; some of them have been experimentally verified to an accuracy of better than one part in 10 billion. However, the theory also predicts that certain quantities (such as the mass of the electron) are infinite. This defect can be repaired by using a somewhat contrived mathematical procedure called renormalization. In effect, the troublesome infinities are "subtracted out." However, the renormalization method doesn't seem to be a mathematically logical one. It is used solely for the purpose of making the theory yield the correct results.

Forces and Particles

Finally, sometimes there appears to be no obvious reason why things are as they are. For example, there are four known forces. These are gravity, electromagnetism, and the strong and weak nuclear forces. The strong force is the one that binds protons and neutrons together in atomic nuclei, and the weak force is responsible for certain kinds of radioactive decays. But why are there four forces in the universe, and not two or three or seven? No one knows. Furthermore, there is a huge disparity between the strengths of the various different forces. For example, the strong force is 10^{39} (one thousand trillion trillion trillion) times stronger than gravity. The reason we don't feel this force is that it has a very short range and acts only on the nuclear level. Why should the difference between two basic forces be so large? Again, no one knows. Could a universe in which there were more or fewer forces, or which had forces closer to each other in strength exist? Once again, no one knows.

There are twelve fundamental particles of matter. Only three of these are constituents of ordinary matter, so perhaps it would be simplest if I were to describe these particles first before enumerating the others. These three particles are the electron, the up quark, and the down quark. Here "up" and "down" are just names; they have nothing to do with directions in space. If the quarks were called "Horace" and "Brigette" instead, nothing would be changed. The up and down quarks are components of the proton and the neutron. A proton is made of two up quarks and a down quark, and a neutron consists of one up and two downs. The up quark has a positive electric charge, while the down quark has a smaller negative charge. In the neutron these charges cancel one another out, giving a net electric charge of zero. Antiparticles, such as the positron and the antiup and antidown quarks, are not counted separately.

This family of particles has one other member, the neutrino. Nowadays, this particle is often referred to as the "electron neutrino." As we shall soon see, several different kinds of neutrinos exist. The electron neutrino is not a constituent of matter, but it is often created in nuclear reactions. So it also is a particle that is found in the world around us. It is not one of those particles that has to be created in the laboratory.

The up quark weighs about 8 percent less than an electron, while the down quark weighs 37 percent more. You might think that, as constituents of protons and neutrons, they would have to be much heavier than that. After all, a proton is 1,836 times heavier than an electron. The reason the quarks do not need to be heavy is that most of the mass of a proton or neutron comes from the energy of the interactions between quarks. Einstein's equation $E = mc^2$ tells us that this energy has an equivalent mass.

The sizes of the masses of the electron and the up and down quarks immediately raise the question of why these particles should have the particular masses they do. Could the up quark weigh 12 percent less than the electron in another universe? Could the down quark weigh 30 percent more? For that matter, why does the electron have the particular mass and electric charge that it does? Would it violate any fundamental laws of physics if it weighed half as much and had twice the charge? I suspect that, by now, you've begun to anticipate what the answers to these questions are going to be. Yes, the answer is that no one knows.

There are two other families of particles. The second consists of the muon, the muon neutrino, and two more quarks, named "strange" and "charm." The muon, which derives its name from the Greek letter mu, can best be described as a "heavy electron." Its properties are similar to those of an electron, but it is 207 times heavier. The muon is sometimes seen in cosmic rays, but is not a component of ordinary matter. The strange and charm quarks are constituents of various kinds of heavy particles that are produced in high-energy particle accelerators.

The same pattern is repeated in the third family, which is made up of the tauon (named after the Greek letter tau), the tauon neutrino, and the bottom and top quarks. The members of this family are the heaviest of all. The bottom quark is nearly a thousand times heavier than an electron, while the top quark weighs more than 30,000 times as much.

The same questions can be asked about the second and third families of particles that were asked about the first. For example, why *is* the top quark so heavy compared to all of the other particles? Could the fundamental laws of physics allow it to have a different mass in another universe? For that matter; why are there exactly three families of particles? Why aren't there one or two or six? If only the first family existed, nothing in the world around us would change. If there were more than three, the additional particles would be seen only in the laboratory. Scientists do know, by the way, that there is no fourth family. Theoretical calculations indicate that if there were one, the universe would have expanded out of the big bang at a different rate.

A truly complete theory of the fundamental forces and particles would answer questions like these. It would either tell us that there is some reason why the fundamental forces and particles have the properties they do, or it would explain how these properties could vary from one universe to another. Until such a fundamental theory is found, scientists will not be able to explain why the universe is the way it is. For physicists, at any rate, this is deeply unsatisfying.

Superstring Theory to the Rescue?

Finding a fundamental theory of the particles and forces is not an easy task. Such a theory would have to explain all four of the forces. At present, we have a situation in which quantum mechanics is used to explain the electromagnetic, strong, and weak forces, while general relativity is the theory that explains gravity. But, as we have seen, general relativity and quantum mechanics are mutually inconsistent. Attempts have been made to combine the two in some reasonably straightforward way, but these attempts have failed.

There is one theory—or rather, class of theories—that might be able to accomplish this task. That is superstring theory, which automatically incorporates gravity along with the other forces. However, superstring theory is still in an early stage of development. The scientists who work with the theory have not yet found a way to make it yield a single experimentally testable prediction. It is not even possible to be sure that superstring theory is correct, and no one knows whether or not it will turn out to be a final theory. Nevertheless, the theory has generated a great deal of excitement among physicists, and intensive theoretical research is being conducted at numerous different locations around the world.

During the 1980s a great deal of criticism was leveled at superstring theory by a number of well-known physicists who questioned whether superstring theory was the best approach. These physicists were especially troubled by the fact that the theory seemed to be able to yield no testable predictions. For example, when he was interviewed about superstring theory, Richard Feynman expressed the opinion that the ideas on which the theory were based were "crazy," and added, "I do feel strongly that this is nonsense!" Feynman then went on to talk about the inability of superstring theory to make predictions in its current state of development, making it necessary to make guesses about the details of the theory. "But then they continue to say that it looks like a promising theory," Feynman said, "in spite of the fact that they have to add all these guesses."

Nobel Prize–winning physicist Sheldon Glashow was even harsher, calling superstring theory a "contagious disease" advocated by "kooky fanatics." Like Feynman, Glashow harped on the fact that superstring theory had been unable to make any testable predictions, and compared it to medieval theology. Superstring theory, he said, was full of highly abstract mathematical arguments which, as far as he could tell, had little relation to reality. The superstring theorists, he charged, were like the medieval scholars who argued about how many angels could dance on the head of a pin.

During the 1990s criticisms of superstring theory became more muted. There were several reasons for this. First, it became apparent that the attempts to combine general relativity with quantum mechanics in other ways had all failed. Superstring theory, it appeared, might be the only possible route to success. Second, the superstring theorists continued to make significant progress. They did not find ways to connect their theory with experiment. However, as more of the details of superstring theory were filled in, it began to look more compelling and logical.

Finally, its adherents were no longer calling it a "theory of everything." This term had been invented amid the initial euphoria about superstring theory that existed in the 1980s. It was often said that a theory that explained all of the fundamental particles and forces would be a "final theory" or a "theory of everything." The latter term wasn't meant to imply that it would explain all physical phenomena, only that the theory would lead to an understanding of the fundamental constituents of the universe. Calling superstring theory a "theory of everything" probably aroused skepticism and sometimes ire among the scientists who had worked long and hard to understand the universe by other methods. In other words, the term was poor propaganda.

Superstring theorists have become somewhat more modest today. One of the results is that other scientists are more willing to listen to their claims. Now many of them readily admit that they don't know whether superstring theory will turn out to be a final theory or not. There is always the possibility that superstrings might be explained in terms of ideas that are even more fundamental. It is even possible that a final theory will never be discovered, that theories describing physical reality will turn out to be something like the layers of an onion, and that all we will ever be able to do is peel back one layer after another. Under such circumstances, scientists' understanding of the physical universe would gradually become better, but attempts to gain a complete understanding would never come to an end.

Before I discuss superstring theory any further it will be necessary to jump back in time and explain how and why the idea of extra dimensions of space were first introduced into physics. You may recall that I previously noted that superstring theory is based on the idea that there are nine spatial dimensions, and that six of these are compacted, or rolled up. It would be a good idea to take a close look at the manner in which this rather strange idea originally arose.

Adding a Dimension

After Einstein published his general theory of relativity in 1916, physicists began to wonder if his theory could somehow be combined with the theory of electromagnetism that had been developed by the Scottish physicist James Clerk Maxwell around the middle of the nineteenth century. Maxwell's theory, which had given a unified explanation of electrical and magnetic forces had led to significant advances in physics, and had had practical implications as well. For example, Maxwell's theory had predicted the existence of radio waves. This prediction had been experimentally confirmed by the German physicist Heinrich Hertz, and the invention of radio communication soon followed. Thus there was every reason to believe that if gravity and electromagnetism could be combined in a unified theory, then new and unexpected phenomena might be discovered.

In those days, no one worried about including the strong and weak nuclear forces in a combined theory. Years were to pass before these would be discovered. In 1916, most physicists believed that gravity and electromagnetism were the only fundamental forces that existed in nature and that, in principle, they were capable of explaining everything.

Einstein's theory of general relativity is a four-dimensional theory. The four dimensions are the familiar three dimensions of space and one of time. The theory was formulated in four dimensions because, in relativity, space and time cannot be considered separately. They interact with one another in various different ways. This is the reason why physicists often speak of *spacetime* rather than of "space and time."

In 1919, the Polish mathematician Theodor Kaluza began to wonder what would happen if Einstein's theory were formulated in five dimensions rather than the usual four. When he worked out the mathematics of the theory, he made an astonishing discovery. When an extra dimension was added, two sets of equations appeared. One set was, naturally, Einstein's equations of gravity. The other set was nothing other than Maxwell's equations of electromagnetism. In the five-dimensional world, there was only one force: gravity. But if this five-dimensional gravity was viewed from a four-dimensional perspective, this single force seemed to have two components.

At the time that Kaluza made this discovery, he was only a *privatdozent* (a kind of assistant professor), and he had established no reputation as a scientist. In those days, unknowns did not send their scientific papers directly to journals. They were required to submit their papers to well-known scientists, who then recommended the papers for publication if they considered them to be of value. Since Kaluza's work was based on Einstein's theory, he naturally sent his paper to Einstein.

At first Einstein was intrigued by the idea. He wrote to Kaluza that the idea that unification might be achieved in five dimensions had never occurred to him. But then Einstein began to have second thoughts. He wrote to Kaluza again, saying that he didn't find the arguments in Kaluza's paper to be convincing enough, suggesting that more work ought to be done before it was published. Einstein also suggested that some attempt should be made to suggest experiments that might confirm or falsify the theory.

Nothing much happened for two years. Then, in 1921, Einstein changed his mind and wrote to Kaluza once more, saying that he was recommending the paper for publication after all. The reason for Einstein's change of heart may have been the publication of a unified theory by the German mathematician Hermann Weyl that Einstein found even more unconvincing than Kaluza's.

There were some problems with Kaluza's theory. For example, it could not explain quantum phenomena. And it was not clear whether the extra dimension should be thought of as something real, or whether it was only a kind of mathematical fiction. To make matters worse, it was not clear how the defects in the theory could be remedied.

The next step was taken by the Swedish physicist Oskar Klein in 1926, who suggested a possible reason why Kaluza's extra dimension was not observed. It might be compacted to a size much smaller than that of an atomic nucleus. The reason we did not know that the fifth dimension was there was that it was too tiny to see.

Klein's solution was an elegant one, but problems remained. When the implications of Kaluza's theory were worked out in detail, the theory yielded predictions that were at variance with experimentally known facts. For example, the theory couldn't successfully describe the behavior of a simple particle such as the electron. Attempts to insert electrons into the theory led to the prediction that the ratio of an electron's charge to its mass was greatly different from measured values.

Einstein continued for some time to attempt to see if the idea of five dimensions could be made to work. However, other physicists quickly lost interest. They did so partly because Kaluza's theory apparently couldn't be reconciled with experimentally known facts, and partly because so many of them had become caught up in working out the implications of the newly discovered theory of quantum mechanics. When, during the 1930s, it began to become apparent that there were four fundamental forces in nature, not two, Kaluza's theory was forgotten. With the notable exception of Einstein, most physicists felt there wasn't much point in trying to unify two forces while the other two were left out.

Force, Matter, and SUSY

There are other particles besides the twelve particles of matter that I described earlier. These are also particles of force, such as the photon. Yes, photons are particles of light, but they play another important role. They also transmit the force that causes electrically charged particles to attract and repel one another. For example, two negatively charged electrons repel each other because they constantly exchange photons. A photon will be emitted by one and absorbed by the other, driving the two electrons apart.

An analogy should make this idea somewhat clearer: Imagine that two ice-skaters are standing on a frozen lake and that they throw a heavy ball back and forth. When the first skater throws the ball, the recoil will push him backwards, and when the second skater catches the ball, he will be thrust backwards, too. The analogy breaks down if one tries to use it to explain why a positively charged particle such as a proton and a negatively charged electron attract each other. But then, all analogies break down at some point. This is inevitable when one compares one thing to another that is very much unlike it.

Particles of matter such as electrons and quarks are called *fermions*, after the Italian physicist Enrico Fermi. Force particles are called *bosons* after the Indian physicist Satyendra Nath Bose. *Gluons*, which are responsible for the strong force that binds quarks into neutrons and protons and neutrons and protons into nuclei, are bosons. So are the particles responsible for the weak force and the as yet undiscovered graviton, which presumably carries the force of gravity. There are eight different kinds of gluons in all, and the weak-force particles, called W^+, W^-, and Z^0 are three in number. Here the symbols $+$, $-$, and 0 indicate that the particles are positively charged, negatively charged, and neutral.

The physicists who work in particle physics have found that the particles seen in nature exhibit different kinds of symmetries. These are not physical symmetries like the symmetry of snowflakes. They are abstract and mathematical, and they relate the properties of different kinds of particles. The discovery of mathematical symmetries has been quite important in the field of particle physics because they tell physicists what kinds of theories they should construct.

Thus we have two different kinds of particles, fermions and bosons, particles of matter and particles of force. It seems natural to wonder if the two kinds of particles might not be related in some way. Well, it turns out that if a particular kind of symmetry, called *supersymmetry* or SUSY, is valid in nature, then they are. You might wonder if this fact has any great importance. After all, supersymmetry is just a mathematical idea. Well, it turns out that it is a very important idea. Supersymmetry is what makes superstring theory work. The term "superstring" is nothing other than a shortened version of "supersymmetric string."

The idea that nature is supersymmetric is a highly plausible one. But there is not yet any proof of the idea. However, the theory that it exists can be tested. If supersymmetry does exist, then there must be a class of as yet undiscovered particles, which are called *sparticles*. Sparticles are expected to have very large masses, and twentieth-century particle accelerators were not powerful enough to produce them. However, new, more powerful accelerators were being constructed at the beginning of the twenty-first century, and it is possible that sparticles might be seen by around 2005. If they are discovered, this will not prove that superstring theory is correct. However, it will demonstrate that one of the assumptions on which it is based is valid.

An Accelerator as Large as the Universe

Superstring theory has been called "a part of twenty-first-century physics that fell by chance into the twentieth century." Although it seems to be a promising theory, it is plagued by two major difficulties. As I previously noted, it has proved to be extremely difficult to relate superstring theory to experiment. In 2001, no one was yet able to suggest any experiments that might be performed that would indicate whether it is a correct theory or not. Second, the theory is mathematically difficult. It is so difficult that no one yet knows what the exact equations of superstring theory are. At best, superstring theorists are only able to find approximate solutions to approximate equations.

It is not likely that superstrings will ever be observed. They are far too small. The smaller an object is, the more powerful particle accelerators have to be in order to observe it. Recall that the characteristic size of superstrings is thought to be 10^{-33} centimeters. This is approximately 100 million million times smaller than an atomic nucleus. It has been calculated that a particle accelerator the size of our galaxy, possibly the size of the entire universe, would be needed to directly observe superstrings. By comparison, the largest accelerator so far conceived, the Superconducting Supercollider, was to have been 54 miles in diameter. This accelerator was never built. Funding was canceled by the U.S. Congress shortly after the project got under way.

One of the mathematical complications of superstring theory stems from the fact that the extra spatial dimensions can be intertwined with one another in tens of thousands of different ways, and no one knows which particular configuration describes our universe. Furthermore, there is not just one superstring theory, but five different ones (as we shall soon see, this has turned out to be a blessing, not an evil). This may give you a glimmer of some of the difficulties that the mathematicians and physicists who work with the theory face.

You might think that working with such a theory would be a hopeless task. Nevertheless, the theory has yielded some results. For example, it has suggested a possible solution of why there are three families of matter particles. It has been shown that there is a relationship between the number of families and the way in which the extra spatial dimensions are intertwined. No one knows why the number must be three, rather than, say 4 or 25 or 480. But then, perhaps there isn't a reason. Perhaps different numbers of families do exist in other universes.

Superstring theorists have also discovered that the theory contains a framework for predicting what the masses of the various different particles should be. At this point, theorists are far from being able to predict the mass of any particle. But at least they know there are ways that this might eventually be done. This is a more important result than you might think. Other theories in physics do not provide *any* method for calculating masses. The masses have to be found by experiment and then be inserted into the theories.

I have alluded to the idea that superstring theory could eventually provide us with a theory of all of the forces. In effect, it would combine quantum mechanics with general relativity. One of the most striking things about superstring theory is that gravity does not have to be "put into" the theory. Gravity is something that is included automatically. One might say that, in superstring theory, there is just one force, not four. However, this force exists in a ten-dimensional world, and when it is viewed from our four-dimensional perspective, it takes the form of four different forces.

Since superstring theory does include gravity, it may well eventually be possible to apply it to problems in cosmology. For example, no one really knows the origin of the force that is causing the expansion of the universe to accelerate. But there is reason to think that superstring theory may eventually provide an answer. Superstring theory is also likely to provide scientists with new insights into processes that took place in the very early universe before the beginning of the inflationary expansion. Superstring theory may even tell us why there was a big bang in the first place.

Although some progress has been made, there remain fundamental questions that have not yet been answered. For example, it is not clear whether superstrings are objects that exist in space and time, or whether space and time are made of superstrings. And then there is the question of the fundamental principles on which superstring theory depends. Recall that Einstein discovered his special and general theories of relativity by working out the implications of certain basic principles. This is *not* the way superstring theory was discovered. There is good reason to expect that there exist reasonably simple basic ideas from which the equations of superstring theory could be derived. Most of the other theories in physics were developed this way. However, no one has any idea what the principles underlying superstring theory are. As a result, theorists have been forced to grope their way through masses of complicated equations, laboriously wringing out results. This isn't the way that one would like to do physics. Perhaps if superstring theory had remained undiscovered until perhaps the middle of the twenty-first century, scientists would find themselves working out the equations in some clear, logical way. Unfortunately, this is not the case. As a result, working on the theory has remained extraordinarily difficult.

M Theory

The fact that there were five distinct superstring theories was always something of an embarrassment. Presumably there should be one fundamental theory that describes our universe, not several. Deciding which of the five described our universe was not an easy task. Theorists had nothing to guide them. Apparently their only choice was to study all five theories in the hope that someone would eventually discover which of them was correct.

Then, in 1995, the situation suddenly changed. At a superstring conference held in March of that year at the University of Southern California, Princeton physicist Edward Witten announced his discovery that there was evidence that the five theories were closely related, even though they looked very different from one another. They described the same physics in different ways. Furthermore, all five of the ten-dimensional superstring theories were related to a single eleven-dimensional theory that has come to be called *M theory*. The five theories were not as unlike one another as they had seemed. They were only five different windows on multidimensional reality.

The discovery of the relationships among the different theories suddenly made doing certain calculations in superstring theory easier. If a calculation was too difficult to perform using one theory, the same calculation might be easier in one of the other theories. Of course, I am speaking in relative terms here. In the context of superstring theory, "easier" often means "still very difficult but not totally impossible." Witten's discovery showed that calculations made in any of the five theories could be recast in a somewhat different form.

It has since been discovered that M theory is also related to a sixth theory, known as *supergravity*. Supergravity is an eleven-dimensional theory of particle physics that was developed during the late 1970s and early 1980s. It is supersymmetric, but it is not a string theory. At first supergravity appeared to be a promising approach. However, it was eventually abandoned because it did not seem to be able to predict all of the particles that physicists observed. The discovery that it was related to superstring theory was somewhat surprising; after all, the basic premises of the supergravity and superstring theories were different.

Witten's announcement set off what has become known as the "second superstring revolution" (the first revolution took place in the mid-1980s when it was shown that superstring theories could be made to work), and has led to the development of a host of new ideas. For example, M theory contains vibrating strings just as superstring theory does. But it predicts the existence of a host of other kinds of objects as well. M theory suggests that there also exist vibrating two-dimensional membranes, vibrating three-dimensional objects, and so on. These objects are called *branes*. For example, a vibrating five-dimensional object is a five-brane. One that had nine dimensions (the maximum) would be a nine-brane.

A lot about M theory is not yet understood. It would not be inaccurate to say that physicists do not yet know what M theory is. They only know that it relates six other theories, and that it apparently has some features, such as the existence of vibrating multiple-dimension objects, that the other theories do not possess. Witten and other physicists have argued that, at low energies (that is, when the particles being described are not very energetic), supergravity is a good approximation of M theory. However, no one yet has any idea what form M theory takes at high energies. This question is currently being investigated theoretically, but it is impossible to say how long it will take before a more complete theoretical understanding is obtained.

When Witten coined the name M theory, he didn't say what the M stood for. It can be taken to mean membrane theory, since membranes are one of its ingredients. It is possible that this or some other name will stick to the theory in the future. But, for now, the "M" in M theory is like the "S" in Harry S. Truman. It doesn't stand for anything.

So Why Are Things as They Are?

So far, scientists have only a few hints. Superstring and M theory has not advanced to the point where it is possible to predict the masses or electric charges of the fundamental particles. Nor have these theories told us why the four forces have the strengths they have. They have discovered a possible theoretical reason why there are three families of matter particles, but they don't know whether or not this number must be three. They are not yet sure what properties of our universe could not be otherwise, or how the laws of physics could vary from universe to universe—if other universes really exist.

They can't even be sure that superstring theory is a correct theory. If supersymmetric particles are discovered, this will provide evidence that supersymmetry exists in nature, but it will not be evidence for the existence of superstrings or of extra dimensions. It is even conceivable that we will discover that nature is *not* supersymmetric. No one thinks this is very likely. The basic laws of physics exhibit a number of other mathematical symmetries, and it is hard to believe that nature would make use of them all while neglecting supersymmetry. But it is possible. And if it turned out that supersymmetry did not exist, then superstring theory would have to be abandoned.

It may appear that little progress has been made toward answering fundamental questions about the nature of our universe. Perhaps this is true. However, a method for attaining this kind of understanding has been discovered. And this is sufficient to motivate numerous physicists around the world to continue their investigations of superstring and M theory.

6 | *What Happened before the Beginning?*

In his *Confessions* St. Augustine speaks of those who asked, "What was God doing before He made the heavens and the earth?" It is sometimes said that Augustine answered, "Preparing hell for those who ask such questions." But in reality, he wrote that he would not give so frivolous an answer, and he made a serious attempt to find a reply.

Questions about the beginning had not been much of a problem for the classical Greek philosophers, who generally believed that the world had always existed. Even those such as Plato and Aristotle, who did not advocate the idea of eternal cycles, gave little or no thought to the idea of a beginning. Plato does speak of a creation by a demiurge in his dialogue *Timaeus*, but Plato's demiurge is not a Creator, he only gives form to matter that was already there.

However, for Augustine, the idea of a Creation did raise some baffling questions. Why had God created the world at a particular point in time? Why hadn't he created it sooner? Presumably, God had always existed. Did this imply that he had allowed an eternity to pass before the Creation? Since Augustine was steeped in Greek philosophy, he was well aware that the classical authors would be of no help. He realized he had to find an original solution to the problem. And of course he did.

Augustine's answer to the question, "What was God doing before he made the heaven and the earth?" was that there had been no "before." Both time and space had come into being when the world was created. God dwelled in Eternity, which was not a kind of time. On the contrary, God remained outside of time. To ask what he had been doing before he created the world was a meaningless question.

When asked, "What happened before the big bang?" cosmologists have sometimes given an answer similar to Augustine's. There was no "before," they have often said. Time and space were created in the big bang. But of course this answer is only a guess. Without a theory or quantum gravity, it is not possible to understand what was happening at the moment of creation, or during the first 10^{-43} seconds. Though it may seem likely that space and time were created together, there is plenty of room for speculation.

Speculation that the universe might have existed in some form before the big bang has been carried on for quite some time. In 1922, the Russian mathematician Alexander Friedmann proposed an "oscillating universe" hypothesis. According to this proposal, which was expounded in detail by the American physicist Richard C. Tolman ten years later, the universe might go through one cycle of expansion and contraction after another. A contracting universe, Friedman and Tolman suggested, might not be destroyed in a big crunch. Perhaps it could somehow "bounce" and re-expand in a new big bang.

Yes, it is currently believed that the universe will probably go on expanding forever. However, cosmology is not a static science. Ideas that seemed well established have frequently been overturned in the past. Thus, though a future contracting universe seems unlikely, it is probably worthwhile to try to see what its eventual fate would be.

Einstein's theory of general relativity can no more describe the universe in the final stages of a big crunch than it can the earliest moments of the big bang. General relativity says that a contracting universe must eventually collapse into a *singularity*, a point-sized region of infinite matter density, infinite energy, and infinite temperature. If the equations of general relativity are correct, then according to theorems worked out by Roger Penrose and Stephen Hawking this is unavoidable.

This is a good reason for believing that general relativity cannot describe the final moments of a contracting universe. When infinite quantities are encountered in the equations of physics, it is always a sign that something has gone terribly wrong, and there is reason for thinking that the theory that one is using has broken down. Thus no one can say that a universe that goes through endless cycles is impossible. There is a small chance that the ancient Stoics were right when they said that the world was destined to be destroyed and re-created endlessly.

Over the ensuing decades, the oscillating universe theory was debated, and new versions were proposed. Some physicists worried about the possibility that starlight might accumulate from cycle to cycle. If it did, they pointed out, then it was possible to place a limit on the number of cycles that could have taken place in the past. Meanwhile, other physicists proposed versions of the idea in which that problem was avoided. John Archibald Wheeler, for example, suggested that perhaps the very laws of physics changed when the universe collapsed in a big crunch, and then re-exploded in a new big bang.

However, all of the scientists who speculated about the matter were hampered by no theory to explain how or why a bounce took place. They were in a position analogous to that of someone trying to find his way through a desert without a map. There was no way of telling whether any of the ideas that were proposed had any chance of being true or not. The oscillating universe idea was extremely speculative when it was first proposed in 1922, and it was just as speculative fifty years later.

Then, in 1984, matters suddenly changed. Two Japanese physicists, Keiji Kikkawa and Masami Yamasaki, discovered that superstring theory implied that all the dimensions of space had a minimum size equal to the Planck length of 10^{-33} centimeters. If the universe contracted down to the Planck length, it could become no smaller. It would bounce and begin expanding again. According to superstring theory, the geometry of space changes in such a way at very small distances that this is inevitable.

Black Holes as Seeds of New Universes

Kikkawa and Yamasaki's discovery was a fascinating result that deepened scientists' understanding of superstring theory. However, the fact that a contracting universe might eventually bounce is probably not too relevant to our universe, which is likely to go on expanding forever. Nor is there any reason to think that our universe was born out of the big crunch of a universe that preceded it. However, this really hasn't put a damper on speculation about what might have happened "before" our universe was created.

I have put the word "before" in quotation marks because, if space and time did come into existence at the moment of the big bang, there were no such things as moments in time before this event. The creation of our universe could still have been caused by events that took place in some other "parent" universe. During the 1980s, physicists who specialized in general relativity began to realize that it was perfectly possible that new universes might be born inside black holes.

Black holes are the collapsed remnants of massive dead stars. They exert gravitational forces that are so strong that nothing, not even light, can escape from them. Most stars die relatively quiet deaths when they exhaust their nuclear fuel. For example, our sun will eventually become a *white dwarf*, a star that continues to glow due to its residual heat after the nuclear reactions in its core have ceased. However, very large stars experience more violent deaths. If the mass of a star is more than about three or four times the mass of our sun at the end of its life, then it will undergo a sudden massive collapse, followed by a supernova explosion. The collapsing matter will then be condensed further by gravity. If it becomes dense enough, a black hole will be created.

General relativity tells us that matter that falls into a black hole must collapse into a region of infinite matter density called a singularity. Such a singularity would be the exact analogue of singularities at the beginning of the big bang or at the end of the big crunch. As we have seen, there are good reasons to believe that such infinite density states do not exist. In particular, superstring theory implies that there is a minimum size, and that nothing can be confined to a region smaller than this.

Nevertheless, conditions near the center of a black hole are more extreme than anything else that can be imagined.

If the general theory of relativity is correct, it is perfectly possible that matter that has fallen into a black hole could pass through a *wormhole* into a newly created region of space and time. A wormhole is a "tunnel" in spacetime that can connect two distant parts of the universe, or two different universes. If wormholes exist, they are probably submicroscopic objects that exist for very short periods of time. If the matter in a black hole did pass through a wormhole, the wormhole would quickly pinch off, and a new *baby universe* ("baby universe" sounds like an attempt to be cute; however, it has become accepted scientific terminology) would be created. This baby universe might or might not undergo an inflationary expansion. If it did, a full-fledged universe would be created.

This is not something that scientists could actually observe. Since no light or any other kind of radiation can escape from a black hole, it is impossible to see inside it. And, according to general relativity, we would not be aware of any change in a black hole's mass if matter passed through a wormhole inside it. An observer outside the black hole would observe no changes in the gravitational attraction exerted by the hole.

It is useless to ask "where" a newly formed universe was located. Once the wormhole vanished, all communication with the parent universe would be broken off and the baby universe would exist in a spacetime of its own. In some respects, the creation of a new universe in this manner would resemble the creation of a universe in a quantum fluctuation. However, there are important differences. The creation of a universe in this manner depends upon the existence of a parent universe that is capable of producing black holes. It is not like the "creation out of nothing" that has been postulated by some advocates of the quantum fluctuation hypothesis.

It is possible that black holes might contain the seeds of new universes, and that our universe was born in just such a way. However, there seems to be no way to test this particular hypothesis. If a group of scientists ever ventured into a black hole, they could never return to tell us what they had observed. It would be perfectly feasible to do this, by the way, if we had some way to journey to a suitable black hole. Black holes created in supernova explosions wouldn't do, because their gravity creates tidal forces that would tear apart any traveler who ventured inside. However, the cores of most galaxies, including our own, contain massive black holes that have masses hundreds of millions or billions of times greater than our sun. Since these black holes are much larger than those that form in supernova explosions, the tidal forces at their surfaces are less intense, and it should be perfectly possible to enter one without experiencing any catastrophic events. However, we can probably assume that it is not likely this will ever be attempted. Scientists like to be able to communicate their discoveries to other scientists, after all. And members of a group of scientists trapped inside a black hole would only be able to converse among themselves.

Evolving Universes

The fact that the black hole hypothesis cannot be tested has not put a damper on further speculation. The American physicist Lee Smolin, for example, has suggested that if black holes do indeed contain the seeds of new universes, then universes might actually be able to evolve. He points out that under such circumstances the universes that would have the most "offspring" would be those that contained the greatest number of black holes. If all universes did not have exactly the same properties, some would have more progeny than others. A very large universe that contained a lot of black holes would produce more baby universes than a small one with few black holes.

Smolin now makes an additional assumption. He hypothesizes that the laws of nature that govern a daughter universe might be very slightly different from the laws in its parent. There is no particular reason to think that this would be the case. However, there are no good arguments against the idea, either. No one knows whether or not baby universes do indeed come from black holes, or what laws might govern their creation. As a result, one is free to assume whatever one wants and then try to see what the consequences would be.

So as far as anyone knows, other universes might not have natural laws identical to those of our own. The superstring theorists hope that they will eventually discover why the laws of physics are as they are. But they have not done so yet, and they are not likely to for some time to come. And that there are thousands of different possible versions of superstring theory opens up the possibility that different versions might be valid in different universes. It is conceivable that in some universes the strength of the force of gravity might be greater than it is in ours, or alternatively, that it might be less. Other universes might contain particles that are similar to those in our universe, but not identical. The masses of the fundamental particles might be different, or the size of the electric charge of an electron might be different. The forces that hold atomic nuclei together might be a little stronger or weaker. The velocity of light might be different. It is possible that universes might vary in a number of different ways.

According to Smolin, if the laws of physics could vary, then this would lead to a kind of natural selection that would act to produce universes that would be efficient at producing black holes and new baby universes. Universes with physical laws such that they produced few offspring would eventually die out. Note that this hypothesis depends on the assumption that there exists a kind of heredity that causes baby universes to be similar to their parents. If the laws of physics varied randomly, or if there were large differences between parent universes and their progeny, natural selection wouldn't work. What is needed is something similar to what happens in the biological world, where evolution proceeds by the accumulation of small mutations. Large mutations are generally harmful and thus not very likely to be passed along to succeeding generations.

Superstring theory provides us with some hints as to how this cosmic heredity might operate. Recall that the various compacted dimensions are entangled with one another in complicated ways. It is conceivable that when a baby universe was born the configuration might change slightly. Naturally, we shouldn't leap to the conclusion that this does indeed happen. Many things about the creation of new universes are not understood. The idea that baby universes are produced by black holes is itself pure speculation. Smolin's proposal may appear far-fetched, but many ideas in physics seemed far-fetched when they were first proposed. And there is one kind of evidence that works in favor of Smolin's theory. Our own universe produces black holes quite frequently. And this is just what we would expect if the hypothesis were correct. It is easy to imagine small changes in the laws of physics that would prevent black hole formation or make it a rare event.

On the other hand, it is possible to conceive of universes in which black hole formation was much more likely than it is in ours. A slightly different "fine-tuning" of the strength of gravitational forces and the forces that operate at the subatomic level could probably produce a universe in which medium-size stars like our sun could evolve into black holes. This seems to imply that if this kind of cosmic evolution is indeed taking place, then it may have a long way to go. Our universe, the only one we know, is far from being optimally adapted. Of course, this may not mean very much. There may be reasons why universes that produce such large numbers of black holes are unlikely.

Smolin's proposal is highly speculative. There are no good reasons for accepting his idea, even provisionally. But when superstring theory and M theory are developed further, they may very well shed some light on what is possible, and what is not possible. At this point, few speculations about the creation of universes can be automatically dismissed.

Chaotic Inflation

The Russian-American physicist Andre Linde has been a major contributor to the development of the inflationary universe theory from the very beginning. During the late 1970s he discovered many of the ideas associated with the theory independently of Alan Guth. It was Linde who developed the "new inflationary universe" theory in 1981, solving certain theoretical problems that arose in Guth's original version of the theory.

Nowadays Linde is perhaps best known for his *chaotic inflationary universe* theory, which he discovered in 1983. Until this time, scientists working with inflationary universe theory had assumed that the universe was initially very hot. It seemed to be necessary to make this assumption if the theory was to work. But no one could explain what caused it to be so hot. Linde's theory solved this problem by doing away with the assumption of a hot pre-inflationary universe.

All versions of the inflationary universe theory depend upon the assumption that quantum fields filled all of space in the period before the inflation began. This idea isn't as abstruse as you might think. Fields are associated with all quantum particles. For example, the charge of an electron creates an electrical field that surrounds it. Gravitational fields surround any particle or collection of particles that have mass, and fields are also associated with the forces that hold nuclei together. Nor does this exhaust the list. Particle physicists believe that there exists something called the *Higgs field* (named after the British physicist Peter Higgs) that is responsible for giving mass to elementary particles. In a sense, it "fattens them up." The Higgs field is thought to be associated with a *Higgs particle*. In 2001, the Higgs particle had not yet been discovered; there were as yet no particle accelerators powerful enough to produce it. However, few physicists doubt that it exists.

Linde's theory began with the assumption that a large variety of different quantum fields existed in the pre-inflationary universe. He assumed also that the strength and character of these fields varied randomly. In one region of the universe, they would have one particular configuration, and in other regions they would be very different. It may seem that Linde was making a lot of assumptions. But he really wasn't. He was simply postulating that conditions in the pre-inflationary universe could have been completely chaotic.

In the chaotic inflationary universe theory, the universe does not originally have to be hot. The existence of chaos is the only requirement. Now, if chaos did reign, different regions of the universe would have been affected differently. In particular, inflationary expansions would have begun in some regions, but not in others, and different parts of the universe would have inflated to different degrees. The regions that didn't expand have presumably remained microscopic, while others became "astronomically"—perhaps some such term as "super-astronomically" might be better here—large. If the theory is correct, we live in one of the regions that once inflated, and the other regions are so far away that neither we nor our distant descendents will ever be able to observe them. No one knows how big our region of the universe is. If the universe is 13 or 14 billion years old, we cannot see farther than 13 or 14 billion light years out into space. There has not been enough time for light to reach us from objects that are farther away. But 13 or 14 billion light years may be a distance that is very small compared to its total extent.

If Linde's theory is correct, then there is no reason why inflationary expansions could not be taking place at numerous different locations today. Inflation could also begin in tiny regions of an existing universe, creating a baby universe that then budded off into another region of space and time. We could no more observe these baby universes than we could the ones that might be created in quantum fluctuations or in black holes. If something like this is happening, new universes can be created without end.

Linde calls this hypothetical process "eternal inflation." Note that if his idea is correct, then our universe need not be one of the first universes to be created. The creation of new universes could be something that has been going on eternally. According to Linde's theory, there is no particular reason to believe that the laws of physics in a baby universe would have to resemble those in the parent universe. For all we know, natural laws could vary randomly from one universe to another. Even the number of dimensions of space and time could change. There could be countless universes that were very much like ours, and countless universes that would seem to us to be very bizarre if we were ever able to observe them.

Even if there is no eternal inflation, and even if conditions were not initially chaotic, there could still be an infinite number of other large "universe-size" regions like our own in the cosmos. There are over fifty other versions of inflationary universe theory. All of them predict inflationary expansions while differing from one another in various details. These theories show that even if conditions were initially uniform, inflationary expansions could have begun in many different places. Over time, the number of expanding regions, which Guth calls "pocket universes," could increase without limit. Over time they would grow in size to dimensions comparable to our own pocket universe.

In the days of the classical Greeks it was believed that Earth and the observable planets *were* the universe. The stars were thought to be embedded in a crystalline sphere that lay at no great distance from Earth. The Copernican theory of a sun-centered solar system and the development of astronomy caused scientists to realize that our sun was only one of numerous stars. The invention of the astronomical telescope multiplied the number of known stars enormously; it quickly became apparent that most stars were too dim to be seen by the naked eye.

The scientific conception of the universe continued to be modified during the twentieth century. At the beginning of the century, it was thought that our Milky Way galaxy *was* the universe. But the development of more powerful telescopes then revealed that many of the "nebula," which had previously been thought to be clouds of gas, were actually galaxies like our own, other "island universes," as they were sometimes called in those days.

Today, astronomers can observe objects that are more than 10 billion light years away, yet it is believed that the observable universe is only a small part of the whole. There appear to be vast reaches of space that are invisible to us because there has not been enough time for light to travel from them to Earth. The very small conceptual universe of the classical Greeks has grown into something enormous.

Thus when matters are viewed in a historical context, the development of the idea of a *multiverse*—a cosmos in which there are many universes—can seem to be a natural next step. Our conception of the universe is again being enlarged, just as it has been so many times in the past.

Universes without End

There appear to be numerous different ways in which new universes *might* be created. All of the mechanisms that have been proposed conform to the known laws of physics. Although all of these theories are speculative, it is not possible to rule any of them out. It almost seems that there are too many different ways to create a new universe. It is hardly credible that all of the proposed mechanisms are operable. Modern cosmology could be said to be suffering not from a lack of ideas, but from too many of them.

These various proposals do seem to have something in common. Time in our universe may have begun in the big bang, but there may always have been a multiverse that has existed eternally. It would be extremely difficult to apply our concept of time to this multiverse. If our universe and certain other universes budded off from different parents, it is not very meaningful to say that one event happened "before" the others. Time in the other universes, after all, would be completely disjoint from ours. If our own universe has a parent, then it has a prehistory. But the prehistory would bear an uncanny resemblance to St. Augustine's eternity, which was supposed to exist outside of time.

And if the laws of physics can indeed vary when new universes are created, there could be some that do not possess time dimensions. It is even conceivable that there exist universes in which there is more than one dimension of time. It is difficult or impossible for us to imagine what a second or third dimension of time would be. However, it is conceivable that universes could exist with multiple time dimensions that contained inhabitants who would not be able to imagine what having a single time dimension would be like.

It is also possible that none of these universe-creation mechanisms could really work. They do not violate any known laws of physics. However, physical laws are yet to be discovered, and the creation of new universes could turn out to be impossible for one reason or another. If we want to know what could conceivably happen, and what could not happen, it will be necessary to learn more about the laws that govern our universe, and possibly other universes.

Superstring Cosmology

At the beginning of the third millennium, a theoretical framework existed that had given physicists few concrete results but seemed to have great promise for the future. This is the theoretical framework discussed in chapter 5, that of superstring and M theory. There is reason to hope that these theories may allow physicists and cosmologists to get a better handle on the question of whether we really do live in a multiverse, and what kinds of properties other universes might have.

When superstring and M theory are more developed, they may tell us a great deal about the early history and possible origin of the universe. At the beginning of the third millennium, superstring cosmology is in its infancy. There have been a few intriguing results, such as the discovery that these theories seem to imply that the universe could never have had less than a certain minimum size. However, most of the other ideas arisen from the theory are just as speculative as those I have been citing.

The first important result obtained was the discovery, made by the American physicists Robert Brandenberger and Cumrun Vafa in the late 1980s, that superstring theory implied that the temperature of the universe was at a maximum when all of its dimensions were about the Planck length, 10^{-33} centimeters. A universe of this size will cool as it expands, and if there exist any forces acting to reduce it to a size smaller than this, it won't contract, but will cool instead. This was yet another argument against the idea suggested by general relativity that the universe began in a state of infinite matter density, energy, and temperature.

Brandenberger and Vafa went on to suggest that the universe was originally in a state where all of its spatial dimensions were about the Planck length. It was hot, but the temperature was not infinite. Then, Brandenberger and Vafa speculated, three of the dimensions began to expand while the others remained compacted. A short time later, those three dimensions underwent an inflationary expansion, producing the universe we see today.

This raised the question of why three dimensions should have become macroscopic, rather than some other number such as two or four or six. The two physicists thought that they had a possible answer for this. It had previously been discovered that theory implied that superstrings could wrap themselves around dimensions of the Planck size, and that this wrapping would prevent the dimensions from expanding. You might think it would follow from this that none of the dimensions should have expanded. However, Brandenberger and Vafa thought this was not the case.

Superstrings are analogous to particles in the sense that there should be both strings and antistrings. When they come into contact with one another, they undergo annihilation, just as particles and antiparticles do. If this happens quickly and often enough, a dimension may become unwrapped and therefore free to expand. Brandenberger then pointed out, that as the number of unwrapped dimensions increases, collisions between strings and antistrings would become progressively more unlikely.

This is analogous to the fact that it is easy to bump into someone if you are walking toward him in a narrow corridor. You are less likely to bump into someone on a wide, two-dimensional sidewalk. Collisions in three-dimensional space are even less likely. Airplanes rarely collide, and when collisions do take place it is generally near airports where air traffic converges. If there were four spatial dimensions, air collisions might never take place. They would be too improbable.

Brandenberger and Vafa calculated that when the number of expanded dimensions reached three, collisions would become infrequent enough that no further unwrapping would take place. The remaining six spatial dimensions would continue to be constricted by superstrings that were wrapped around them. Their picture of the initial moments of the universe was as follows: the high temperature and energy that was present in the early universe caused all of the spatial dimensions to try to expand. Superstrings that were wrapped around them prevented this. Sooner or later, random fluctuations briefly caused some of the dimensions to grow larger than the others. Their larger size made it more likely that the superstrings that wrapped them would collide. Enough string-antistring annihilations then took place to cause three of the dimensions to become unwrapped. These dimensions expanded, and once they began to expand, it was harder for strings to wrap themselves around them and compress them again. Finally, they became large enough that this was impossible.

Brandenberger and Vafa really didn't prove that something like this had to have happened. Like all physicists who work with superstring theory, they had to rely on approximate solutions to approximate equations. Thus their calculations cannot be considered very trustworthy. But their work was important because they at least suggested a possible mechanism for the creation of a universe with three macroscopic spatial dimensions.

Theoretical research on superstring cosmology has continued since Brandenberger and Vafa presented their early universe scenario.

Among the physicists who have done research in the field are Gabriele Veneziano of CERN and his collaborator Maurizio Gasperini of the University of Toronto. Veneziano and Gasperini have created a "pre-big bang" scenario for the early evolution of the universe that is quite different from that of Brandenberger and Vafa. Unlike the latter two physicists, they assumed that our universe began, not as a hot, Planck-size "nugget," but as a small region of a larger universe that was cold, empty, and very large or infinite in extent. As Veneziano points out, it is desirable to try to explain the origin of the universe without having to assume some special initial state. For example, if one assumes that the universe was initially very hot or very chaotic, the reasons for these initial conditions usually remain unexplained. Veneziano and Gasperini's proposal, on the other hand, begins by assuming that the universe was initially in the simplest possible state.

The two physicists then noted that, according to superstring theory, quantum fluctuations would have created irregularities in the pre-big bang universe. According to superstring theory, a kind of inflationary expansion would set in, magnifying the quantum fluctuations enormously. After some time, a tiny three-dimensional region of this universe, perhaps a millimeter in size, would look just like the universe that emerges out of Guth's inflationary expansion. The universe we see today could have originally been a microscopic region of what was originally an essentially infinite universe.

Veneziano and Gasperini's pre-big bang scenario seems to have one advantage over the hypothesis of Brandenberger and Vafa, which leaves the origin of the original Plank-sized nugget unexplained. It does require that the universe be truly vast, since it is postulated that the universe we see grew from a region that was originally submicroscopic. But, as we have seen, scientists' conception of the universe has been changing for centuries. The universe postulated by cosmologists has grown larger with almost every passing decade.

Unlike most of the proposals I have described, the Veneziano-Gasperini pre-big bang scenario could be empirically tested. Some of the magnified quantum fluctuations the scenario predicts could presumably be observed today; for example, as gravitational waves or cosmic magnetic fields. According to the general theory of relativity, a gravitating body that is accelerated produces tiny ripples in spacetime that travel at the speed of light. In 2001, gravitational waves had not yet been detected. However, new, highly sensitive detectors were being constructed. The detection of the gravitational waves, magnetic fields, or other phenomena predicted by the scenario could allow scientists to infer what pre-big bang conditions were like. Alternatively, the data could falsify the hypothesis. As Veneziano says, "In its simplest form, the pre-big bang scenario makes definite predictions that may soon be the cause of its downfall."

Veneziano admits that the specific scenario he and Gasperini proposed may very well turn out to be incorrect. But even if this happens, their accomplishment will remain a significant one. They have shown that with the aid of superstring theory scientists can create testable hypotheses about the origin of the universe. Questions about the manner in which the universe came into being are no longer purely speculative. What were once metaphysical queries have become questions that can be investigated scientifically.

7 | *Are There Parallel Worlds?*

Although the various different scenarios that attempt to describe the origin of the universe have been developed during the last several decades of the twentieth century, some of them are reminiscent of a theory developed by the Greek philosopher Anaximander, who lived in the sixth century B.C.

Anaximander lived in Miletus and was apparently a pupil of Thales. Although all of his writings have been lost, his philosophical doctrines were described by later writers, so we have a reasonably good idea of what he taught. Like Thales, Anaximander believed that there was a single primary substance. However, he departed from Thales by maintaining that it was very unlike any of the substances encountered on Earth. According to Theophastrus, a writer who quoted him, Anaximander described this primary element as follows: "It is neither water nor any other of the so-called elements, but a nature different from them and infinite, from which arise all the heavens and worlds within them."

Anaximander called this substance *apeiron*, which can be translated as "infinite." "Eternal and ageless," he said, it "encompasses all the worlds." He maintained that an eternal motion of the infinite gave birth to innumerable different worlds. These worlds were not eternal; they frequently perished, only to be replaced by the new worlds that were constantly being created.

One probably shouldn't dwell too much on the similarities between this doctrine and theories like Linde's chaotic inflation, which holds that universes like ours are constantly coming into existence in different regions of a vast supercosmos. Anaximander's reasons for developing this idea were quite different from those that motivate contemporary cosmologists. Anaximander sought to discover the one primary substance of which everything was composed, and he apparently didn't feel that Thales's idea that "everything is water" was adequate. However, his theory does show how old the concept of other worlds—or other universes—really is.

The atomist philosophers Leucippus and Democritus later expressed similar ideas. They too believed in a plurality of worlds. However, in the end, the authority of Aristotle prevailed. Aristotle maintained that nothing infinite could exist. Infinite bodies were impossible. Every body had a surface, and anything with a surface had to be finite. Nor were there infinite numbers. There were only potential infinities. An example of a potential infinity would be the positive integers 1, 2, 3. . . . This series of numbers never comes to an end. But, however long one goes on counting, one never encounters a number that is anything but finite.

Aristotle repudiated the idea of a plurality of worlds, proposing that the cosmos was relatively limited in extent. Aristotle's universe consisted of Earth, the planets, and a crystalline sphere that surrounded them. The stars were embedded in this sphere, which was not very large by modern astronomical standards.

Aristotle's conception of the world prevailed in Europe for the next two thousand years. However, the idea of an infinity of worlds didn't die. It is found in the writings of Plutarch, which were penned during the first century. Plutarch tells us that Alexander the Great wept when he heard that there was an infinity of worlds, saying, "Do you not think it a matter worthy of lamentation that when there is such a vast multitude of them, we have not yet conquered one?" There is no reason to believe that Alexander actually said this, of course. During classical times, numerous apocryphal stories about well-known historical figures circulated. If Alexander did say it, or something like it, then perhaps this is yet another indication that he hadn't paid much attention to the teachings of his tutor Aristotle.

One of the reasons Aristotle's doctrine flourished was because the idea of an infinity of worlds was incompatible with Christian doctrine. The Bible spoke only of the creation of one world, Earth, and it would have been considered heretical to postulate the existence of a plurality of inhabited worlds. If such worlds existed, one would have had to assume either that Jesus had preached and been crucified on all of them, or that he had come only to save the inhabitants of Earth. The expression of such heretical ideas would hardly have been tolerated. Once the Inquisition was created, exposition of this kind of idea would have put one in danger of being burned at the stake. As we shall soon see, this is exactly what happened in one case.

An Infinite Universe?

The first person in modern times to suggest that the universe was infinite and that it might contain an infinite numbers of worlds was the English astronomer Thomas Digges, who published a popular account of Copernicus's theory of a sun-centered solar system in 1576. Digges realized that if the theory was correct, then the stars did not have to be positioned on a sphere, as Aristotle had taught. The stars could very well be other suns. According to Digges, Earth and the sun were at the center of the universe, and stars were scattered without limit in every direction in an infinite space.

Digges didn't realize that an infinite space would have no center. A center can only exist in a finite space that has definite boundaries. In an infinite space, every point is a "center" as much as any other. But perhaps this wasn't a significant mistake. Digges did understand the more important point that if the Copernican theory was correct, then the universe could be immeasurably large.

Since Digges lived in England, a Protestant country, he had no cause to fear the Inquisition. In the sixteenth century, the Protestants were no more tolerant of ideas that seemed heretical than the Roman Catholics. However, the Protestant churches were generally subordinate to the secular princes, and they did not have the power of their counterpart, the Roman Catholic Church. No one seems to have suggested that Digges's ideas might be heretical, however, and his ideas appear to have been relatively well known to literate Englishmen. Indeed, it is likely that Shakespeare knew Digges's writings, and that they were the source of some of the astronomical imagery that he used in his plays.

Matters were somewhat different for the Italian philosopher Giordano Bruno, who was accused of heresy even before he began to propound his doctrine of an infinite universe containing an infinite number of inhabited worlds. Bruno joined the Dominican order in 1565, when he was seventeen years old. The monastic life did not agree with him, for his rebelliousness and insubordination to the monastic authorities led to charges of heresy, filed against him in 1576. Bruno fled the monastery, and took up the life of a wandering scholar. During the next two years, he lived in a number of different Italian cities for short times. He then departed for France.

Bruno lived and taught in France, England, Germany, and Switzerland for the next two decades, finally returning to Italy in 1591 when a wealthy Venetian nobleman, Giovanni Moncenigo, offered him a position as a tutor. Bruno must have thought that Venice, which was the most liberal of the Italian states at the time, would be a safe place to live. However, this proved not to be the case. Bruno and Moncenigo did not get along very well, and the nobleman soon became alarmed by the heretical remarks Bruno often made. In 1592, on the advice of his confessor, Moncenigo denounced Bruno to the Inquisition.

Bruno's trial in Venice never reached a conclusion. Before judgment was passed, Bruno was extradited to Rome. After languishing in the prisons of the Roman Inquisition for six years, he was tried again. On February 16, 1600, Bruno was burned at the stake, and it was decreed that his books be publicly burned and their titles placed upon the Index of books that Catholics were forbidden to read.

Sixteenth-century documents indicate that Bruno's inquisitors found some eight heretical propositions in his books. No one knows precisely what these eight propositions were; no document listing them exists today. However, we can reasonably speculate that one of them was Bruno's doctrine that the universe was infinite, and that it contained an infinite number of other worlds, some of which were populated by other human beings.

In his book *De l'infinito universero,* Bruno said:

> Thus is the excellence of God and the greatness of his kingdom made manifest; he is glorified not in one, but in countless suns; not in a single earth, a single world, but in a thousand thousand, I say, in an infinity of worlds.

It is not difficult to see why the ecclesiastical authorities would have found this disturbing, especially when it was expounded along with other heretical doctrines, such as reincarnation, and the idea that Earth—and the universe itself—possessed a soul. It is hard to see how Bruno could have remained alive if he lived anywhere within the reaches of the Inquisition.

Bruno's ideas may have been heretical, but they captured the imagination of the next century. There are numerous references to the doctrine of the plurality of worlds in the writings of John Donne, in Robert Burton's *Anatomy of Melancholy,* in John Milton's *Paradise Lost,* and in the writings of the French essayist Michel de Montaigne. By the end of the seventeenth century, the idea of an infinite number of worlds had become commonplace. In 1734 the English poet Alexander Pope summed it all up in these lines from his poem *Essay on Man*:

> *Through worlds unnumbered though the God be known,*
> *'Tis ours to trace him only in our own.*
> *He, who through vast immensity can pierce,*
> *See worlds on worlds compose one universe.*
> *Observe how system into system run,*
> *What other planets circle other suns,*
> *What varied Being peoples every star,*
> *May tell us why Heaven has made us as we are.*

A Plurality of Pluralities

If the idea of a plurality of worlds was commonplace in Pope's day, the concept of a plurality of universes is just as commonplace today, at least among physicists and cosmologists. Indeed, there are almost too many different ways in which the number of different universes might be infinite.

As we have seen in previous chapters, the many worlds interpretation of quantum mechanics postulates the existence of an infinite number of alternate universes, many of which differ from ours only in minute ways. If this interpretation is correct, numerous copies of ourselves inhabit other worlds, and have lives that are different from ours in numerous different large or small ways. Naturally, there would also be other universes in which human beings never evolved, or in which Earth was lifeless. The only ways in which these universes would be identical to ours would be that they would presumably have the same laws of physics.

As we have also seen, the universe might have arisen as a quantum fluctuation of some kind. A number of different theories are based on this idea. If the universe did come into existence in this way, then there is no reason why this kind of event could not have happened innumerable times. If this idea is correct, we cannot say whether or not the other universes would necessarily have the same laws of physics as our own. As Einstein would have said, physicists do not yet know whether God had any choice when he made the universe. It is possible that physical laws might vary from universe to universe.

This immediately raises the possibility of an infinity of an infinity of universes. Suppose the many worlds interpretation of quantum mechanics is correct, and also that innumerable universes have arisen from quantum fluctuations. Then each one of these innumerable other universes would be duplicated an infinite number of times. The implications of this are truly mind-boggling. It is one thing to speak of an infinity of worlds, as Bruno did, but quite another to speak of an infinity of universes, each of which is infinitely duplicated.

There are other possibilities. It is conceivable that universes may reproduce themselves, that they might be constantly budding off baby universes. This might happen within black holes, or it might happen under other conditions. Finally, there is the possibility, which Alan Guth has studied in detail, that a sufficiently advanced technological civilization might be able to create new universes at will. No one is yet certain whether there really is a multiverse or not. But if there is, we can't rule out the possibility that there exists a multiverse of multiverses.

Gravity, Extra Dimensions, and Parallel Universes

You might think that the various different scenarios would exhaust all the possibilities. They don't. There seems to be another way in which numerous other universes might exist. In 1998, three Stanford University physicists put forward the hypothesis that parallel universes might exist in one of the extra dimensions of space postulated by superstring theory.

When superstring theory was first formulated, it was assumed that all of the extra spatial dimensions would have to be compacted to dimensions of about 10^{-33} centimeters. But during the last decade of the twentieth century, scientists began to realize it was possible that one or more of them was much larger. In 1990 Ignatios Antonisdis of the Ecole Polytechnique in France suggested that some of the dimensions might have sizes of the order of 10^{-17} centimeters. This was still submicroscopic, though much larger than the 10^{-33} centimeters that had previously been assumed.

In recent years, superstring theorists have realized all of the extra dimensions don't even have to be submicroscopic, that it is possible one or more of them could have a macroscopic size. A dimension might be compacted to a size of only about a millimeter (about $\frac{1}{25}$ inch), for example.

Most of the physicists who pursued this line of speculation did so because they were looking for ways in which superstring theory might be tested experimentally. Nima Arkani-Hamed, Savas Dimopoulos, and George Dvali, three physicists who found themselves together at

Stanford University in the late 1990s, took a different tack. Wondering what the cosmological implications of large extra dimensions (large compared to 10^{-33} centimeters) might be, they constructed a speculative theory that implied parallel universes might exist.

Recall that gravity is a very weak force. It is about 10^{39} times weaker than the strong nuclear force, for example. It is far weaker than the other forces as well. For example, when you pick up a paper clip with a small magnet, the magnet exerts more force on the clip than the entire earth. The only reason that gravity is an important force in our universe is that it is the only one that acts over large distances. The strong and weak forces are very strong, but they fall off rapidly at dimensions greater than that of an atomic nucleus. The electromagnetic force, on the other hand, could have a long range if it acted between large electrically charged bodies. However, the matter in our universe—planets, stars, and galaxies—is observed to be electrically neutral. A body that did have excess positive or negative charge would quickly lose that charge. For example, a positive-charged body would attract electrons, which would neutralize the charge.

Arkani-Hamed, Dinopoulos, and Dvali realized that the weakness of gravity could be explained if it was assumed that, unlike the other forces, gravity acted in more than three dimensions. They suggested that perhaps everything that could be seen in our universe, including the strong, weak, and electromagnetic forces, was confined to the usual three macroscopic spatial dimensions, but that gravity "leaked off" into a fourth dimension. This provided a possible explanation for the weakness of gravity, they suggested. Furthermore, there were ways that the idea could be experimentally tested.

Suppose that the extra dimension in which gravity acted had a macroscopic size, say about a millimeter. If this was the case, gravity would behave in its usual familiar way down to about this distance. But at distances of less than a millimeter, its strength would seem to be different than that predicted by Newton's law of gravity. (Newton's law can be used in such a case because its predictions and the predictions of general relativity only differ when there are very strong gravitational fields, or when one is trying to describe the overall structure of the universe.)

When the three physicists proposed their theory in 1998, Newton's law of gravity had never been tested at such small distances. Such experiments are extremely difficult. It is necessary to try to measure the gravitational attraction between objects not much more than a millimeter in size. Such objects must necessarily have small masses, so the gravitational attraction they exert on one another is tiny. Nevertheless, such experiments can be performed, and in the late 1990s, several different physicists began to attempt to carry them out.

I will discuss the results of such experiments shortly. There are some other aspects of Arkani-Hamed's, Dimopoulos's, and Dvali's theory I want to discuss first. The first is the question of why a dimension that has a size of about a millimeter would not be seen. After all, it is possible to see objects of this size with the naked eye; sensitive scientific instruments can detect objects that are far smaller. So why would a macroscopic dimension be invisible to us?

The answer is that it is the electromagnetic force that allows us to see things. Light is electromagnetic radiation, and if the electromagnetic force were confined to the usual three dimensions, then any extra dimension would be invisible to it. Furthermore, there is no other conceivable way of detecting such a dimension except with experiments that depend on the force of gravity. It is invisible to *all* the other forces.

The hypothesis proposed by the three physicists also opens up the possibility that there might exist other universes that are "parallel" to ours in the extra dimension. Indeed, this suggests a possible solution to the problem of the nature of dark matter. If such universes existed, we would not be able to see them; we would be aware only of the gravitational effects of the matter they contained. This raises the possibility that some of the dark matter that astronomers observe is not something in our universe, but is matter in other universes that might be only tiny fractions of a millimeter away.

Decades ago, science-fiction writers invented the idea of parallel universes that existed in the fourth dimension. If Arkani-Hamed's, Dimopoulos's, and Dvali's idea is correct, this could turn out to be literally true. It is not likely that we could ever travel to these other universes. But they could nevertheless exist. It should be remembered, however, that the three scientists did not set out to develop a theory of other universes. Their goal was to try to explain why gravity was such a weak force. The idea of parallel universes arose only as an interesting possibility. It is entirely possible that scientists will eventually discover that dark matter consists entirely of matter present in our three dimensional space. If they do, their results would contradict the parallel universe theory.

When Arkani-Hamed, Dimopoulos, and Dvali proposed their theory in 1998, they suggested that one or more of the extra spatial dimensions might be as large as a millimeter. It has since been shown that they must be smaller than this. In 2000, a group of physicists at the University of Washington in Seattle performed a series of experiments that showed that Newton's law of gravity was accurate down to distances of about two-tenths of a millimeter. This places a limit of the same size on the hypothetical extra macroscopic dimension. The experiment does not disprove the three physicists' theory in any way, but it does show that their conjecture about the size of macroscopic extra dimensions has to be scaled down. And it is entirely possible that further experiments that measure the force of gravity at even smaller distances will yield a positive result.

Other Worlds and the Human Imagination

The idea of an infinite number of other worlds is obviously something that possesses a great deal of appeal to the human imagination. If it did not, then the idea would not have arisen in Classical times, again during the Renaissance, and once more in modern hypotheses of other universes. However, if an idea appears appealing, it does not necessarily follow that it must be true. There have been a lot of captivating ideas that have turned out to be wrong. As the nineteenth-century British biologist T. H. Huxley pointed out, beautiful theories often fall victim to ugly facts.

The various different hypotheses about other universes are all plausible. If our understanding of the implications of quantum mechanics and the conditions under which our universe was created, then it is hard to see how innumerable other universes could *not* exist. On the other hand, there is as yet no evidence to support any of the hypotheses I have described. Are there other universes? We have good reasons for thinking that there very well might be. This is all that one can really say.

8 | *Is There a God?*

It is probably safe to infer that most people would say that this is not the kind of question science is competent to answer. Science, after all, deals with the physical world while religion is concerned with the transcendental. No one expects that God, or a cosmic "made by God" sticker, will be seen with the Hubble space telescope or that signs of his existence will be detected by new, more powerful particle accelerators. It is inconceivable that any empirical test of God's existence will ever be invented. Nor is it likely that any scientist will ever discover a theoretical argument for the existence of God. One would think that the only reasonable answer to the question that constitutes the title of this chapter is: The question of whether or not God exists has nothing whatsoever to do with science.

So why am I writing about this topic at all? I am writing about it because scientists do express opinions on this subject, sometimes giving what seem to be scientific reasons for their belief or unbelief. On the one hand are nonbelievers such as Stephen Hawking and British zoologist Richard Dawkins who claim that modern science is only compatible with atheism. On the other hand are religious scientists and philosophers who maintain that the hand of God can be seen in the design of the universe and in the laws of physics.

In his book *A Brief History of Time,* after describing his no-boundary description of the universe (recall that, according to this hypothesis, the universe originally had four spacelike dimensions and thus had no beginning in time), Hawking wrote:

> The idea that space and time may form a closed surface without boundary also has profound implications for the role of God in the universe. . . . So long as the universe had a beginning, we could suppose it had a creator. But if the universe is really completely self-contained, having no boundary or edge, it would have neither beginning nor end; it would simply be. What place, then, for a creator?

In *River out of Eden,* Richard Dawkins expressed himself somewhat differently:

> If the universe were just electrons and selfish genes, meaningless tragedies like the crashing of a bus are exactly what we should expect, along with equally meaningless *good* [his italics] fortune. Such a universe would be neither evil nor good in intention. It would manifest no intentions of any kind. In a universe of blind physical forces, some people are going to get hurt, and other people are going to get lucky, and you won't find any rhyme or reason in it, nor any justice.

Taken out of context, this statement can be read either as an argument against the idea of a godless universe or one in favor of the idea. However, anyone who reads Dawkins at length soon realizes that he believes that science does tell us that we live in a universe without purpose, without justice, and without God.

For example, in *River out of Eden* Dawkins characterized religious faith as "the great cop-out, the great excuse to evade the need to think and evaluate evidence." He said, "Faith is belief in spite of, even perhaps because of, the lack of evidence." In 1992, during a debate with the Archbishop of York, Dr. John Hapgood, Dawkins said, "The more you understand the significance of evolution, the more you are pushed away from the agnostic position and towards atheism."

Naturally, not all scientists agree that scientific knowledge implies atheism. During the twentieth century the British philosophers Frederick R. Tennant (in *Philosophical Theology*, 1928–1930) and Richard Swinburne (in *The Existence of God*, 1979) resurrected an old argument, known as the teleological argument, or argument from design, maintaining that the order and functioning of the universe and the "fit" between it and the human mind did imply the existence of a creator. Nowadays, similar arguments are often made by religiously inclined philosophers and scientists. Attempts have also been made to show that the findings of modern cosmology imply that the universe was consciously designed. It is pointed out that if the laws of physics were slightly different in any of a number of respects, then life couldn't possibly exist. The claim is then made that a universe containing life is so improbable a thing that the nature of the laws of physics cannot be ascribed to chance.

I will describe some of these arguments in detail shortly. However, I think it might be illuminating if I first gave a short history of the argument from design. This argument is one that has been alternately accepted and rejected. It was used in ancient times, by such scholastic philosophers as St. Thomas Aquinas. It was strongly criticized by Hume and by Kant, but found numerous supporters in the seventeenth, eighteenth, and nineteenth centuries. By the end of the nineteenth century it was thought to have been discredited, and then, as we have seen, it was resurrected once again in the twentieth century.

The Turbulent History of an Idea

Like many others who lived at the time, St. Paul believed that the existence of God could be inferred from the design that was evident in nature. In his epistle to the Romans (Romans 1:20), Paul said, "For since the creation of the world God's invisible qualities, his eternal power and divine nature, have been clearly seen, being understood from what has been made, so that men are without excuse."

There were undoubtedly many who would have disagreed with St. Paul. At the time, the philosophy of Epicureanism was still popular in the Roman world, and Epicurus had maintained that everything in the world had been created when the atoms of which matter was composed spontaneously "swerved" and coalesced into bodies, including inanimate objects, animals, human beings, and gods. The gods who were formed in this way had not created the world, nor did they have any control over it. Like everything else, they were a product of chance.

In *On the Nature of the Gods,* the Roman orator, prosecuting attorney, and philosopher Cicero attempted to refute these ideas, maintaining that evidence of an intelligent designer could be observed in nature. A conscious purpose could be seen in a painting or a statue, Cicero argued. But nature was more perfect than art. Consequently, it exhibited purpose, too. Similarly, sundials and water clocks measured time by design rather than by chance. Even ignorant barbarians should be able to see that the more complex movements of the sun, the stars, and the planets had to be the work of a conscious designer.

Cicero continues, maintaining that it was implausible that the chance collisions of particles postulated by Epicurus could have produced anything as beautiful as the world. According to Cicero, this was like believing that if the letters of the alphabet were thrown on the ground often enough they would sooner or later combine to spell out the *Annals of Ennius* (a narrative poem by the Roman poet Quintus Ennius).

The argument from design was accepted by the thirteenth-century Italian philosopher St. Thomas Aquinas, who presented it as one of his "five proofs" for the existence of God. The argument was also discussed by a number of other medieval thinkers. But in modern times, it began to come under philosophical attack, notably by the eighteenth-century Scottish philosopher David Hume and the eighteenth-century German philosopher Immanuel Kant. According to Hume, the order that appeared in the world could indeed have come about by chance. In such a case, the world would have the appearance of design, but no designer. Furthermore, Hume said, even if an intelligent designer could be inferred, it would not be possible to claim that this designer was the infinitely good and powerful God of Christian faith.

Kant's arguments abut the matter were somewhat more complex. Basically he argued that additional philosophical assumptions were needed to establish the validity of the argument, and that these assumptions were by no means certain. Furthermore, Kant said, the design argument proved only that the form of the world had been caused. It didn't imply that the matter of which the world was composed had been caused, too.

If the argument from design had continued to be nothing but a topic for philosophical debate, I probably wouldn't be writing about it. However, the argument began to take on a different character in the seventeenth, eighteenth, and nineteenth centuries. The newly earned prestige of science caused empirical arguments to become as highly valued as philosophical ones, if not more so. Attempts were therefore made to add an empirical element to the design argument.

A Watch upon the Heath

The empirical arguments reached their culmination in the writings of the English Anglican priest and Utilitarian philosopher William Paley, who published several influential books between 1785 and 1802, including *A View of the Evidence of Christianity* (1794), which was required reading for entrance to Cambridge University until the twentieth century, and *Natural Theology,* in which Paley expounded on the analogy of a watch found upon the heath. The passage that begins the book has often been quoted. I think it is worth quoting it again here, since it is the kind of argument that must have seemed quite conclusive to Paley's readers. The argument is a little long-winded—which wasn't uncommon in his day—but nevertheless worth reading in its entirety (all of the italics in this passage are Paley's):

> In crossing a heath, suppose I pitched my foot against a *stone* and were asked how it came to be there; I might possibly answer that, for any thing I knew to the contrary, it had

lain there for ever: nor would it perhaps be very easy to show the absurdity of this answer. But suppose I had found a *watch* upon the ground, and it should be inquired how the watch happened to be in that place; I should hardly think of the answer which I had before given, that, for any thing I knew, the watch might have always been there. Yet why should not this answer serve for the watch as well as for the stone? why is it not as admissible in the second case, as in the first? For this reason and for no other, viz., that, when we come to inspect the watch, we perceive (what we could not discover in the stone) that its several parts are framed and put together for a purpose, *e.g.* that they are so formed and adjusted as to produce motion, and that motion so regulated as to point out the hour of the day; that, if the different parts had been differently shaped from what they are, of a different size from what they are, or placed after any other manner, or in any other order, than that in which they are placed, either no motion at all would have been carried on in the machine, or none which would have answered the use that is now served by it. To reckon up a few of the plainest of these parts, and of their offices, all tending to one result: —We see a cylindrical box containing a coiled elastic spring, which, by its endeavor to relax itself, turns round the box. We next observe a flexible chain (artificially wrought for the sake of flexure), communicating the action of the spring from the box to the fusee. We then find a series of wheels, the teeth of which catch in, and apply to each other, conducting the motion from the fusee to the balance, and from the balance to the pointer; and at the same time, by the size and shape of those wheels, so regulating the motion, as to terminate in causing an index, by an equable and measured progression, to pass over a given space in a given time. We take notice that the wheels are made of brass in order to keep them from rust; the springs of steel, no other metal being so elastic; that over the face of the watch there is placed a glass, a material employed in no other part of the work, but in the room of

which, if there had been any other than a transparent substance, the hour could not be seen without opening the case.
The mechanism being observed (it requires indeed an examination of the instrument, and perhaps some previous knowledge of the subject, to perceive and understand it; but being
once, as we have said, observed and understood), the infer-
ence, we think is inevitable, that the watch must have had a
maker: that there must have existed, at some time, and at
some place or other, an artificer or artificers who formed it
for the purpose which we find it actually to answer; who
comprehended its construction, and designed its use.

The modern reader is likely to become somewhat impatient with
this, feeling perhaps that Paley is belaboring an obvious point. However, the author of *Natural Theology* has reasons for presenting his argument in such detail. The bulk of his book is devoted to equally detailed
examinations of examples of design in the natural world. Paley, who
was familiar with the science of his day, gives one example of design
after another. He begins with a discussion of the design of the human
eye, maintaining that its structure alone is evidence of the hand of a creator. God, Paley says, could have made it possible to see by divine fiat;
he did not need to create an artfully contrived eye. But the Deity, Paley
goes on, obviously designed the intricacies of the eye as testimony to his
existence.

Paley wasn't the first to argue in such a way. According to his contemporary Xenophon, Socrates had discussed the eye as evidence of
intelligent design. However, Socrates made only a brief argument. He
didn't go into anything like the detail Paley did. Paley was not content
with an examination of the overall design of the eye; he went into as
much detail as he did when speaking about the watch, commenting on
the design of each individual part of the eye.

Paley then goes on to examine other parts of the human body, such
as the stomach, the neck, the spine, the human hand, and even the ten-
dons, claiming that each of them contains such evidence of design that
we have no choice but to conclude there was a divine Designer. Nor
does he confine himself to an examination of the human body. He also

sees evidence of design in the feathers of birds and the tails of fish, the fangs of snakes, the stomach of the camel, the beak of the parrot, and the tongue of the woodpecker. Furthermore, according to Paley the sciences of chemistry and astronomy provide evidences of design in the natural world. The properties of water, he says, also show evidence of design and forethought. There are so many examples of design in the world, he concludes, as to make atheism absurd.

Paley's arguments are not philosophical ones. He was attempting to make use of scientific knowledge in order to demonstrate the existence of God. In 1802, when *Natural Theology* was published, and in the decades that followed, his arguments must have seemed forceful ones. After all, the phenomena that Paley described obviously could not have happened by chance. But chance seemed to be the only alternative to deliberate design by a creator.

Design and Natural Selection

Charles Darwin, who attended Cambridge University as a young man, was quite familiar with Paley's writings. Indeed, he was strongly influenced by Paley's *Natural Theology* when he was a young man. Later Darwin found himself in disagreement with Paley's conclusions. The argument from design was not compatible with Darwin's growing conviction that blind natural selection was the source of all the design observed in nature.

Darwin was not an atheist, and he had no interest in trying to overturn Paley's conclusions about the existence of God. However, he did realize that the existence of design could be explained scientifically, and that it need not be ascribed to a creator. In his book *Origin of Species,* which was published in 1859, Darwin argued that natural selection was quite capable of creating the structures in the human body and in animals that Paley had marveled over. So convincing were Darwin's arguments that, for all practical purposes, the argument from design fell into disuse.

The argument never vanished completely. I recall having heard Paley's analogy of the watch upon the heath in Sunday School when I was a child. Religious writers continued to appeal to it well into the twentieth century. However, the argument did disappear from serious scientific and philosophical debate.

Natural Selection

The idea of natural selection is really very simple. If the number of individuals in a species remains roughly constant, then only some of them will survive and reproduce. This is true of fish and insects that produce large numbers of eggs, and it is true of slow-breeding animals such as the elephant. Darwin himself calculated that a single pair of elephants would have 15 million descendants after only about 750 years. Since elephants have been around for millions of years, the fact that our planet is not overrun with them forces us to conclude that only a fraction of the elephants that are born have descendants.

Some animals die before they reach reproductive age, and some fail to have any offspring. Darwin noted that if there was variation among the individuals that make up a species, then those that did reproduce would pass along their traits to their offspring, while the other individuals would not. In time, this would cause these favorable characteristics to spread throughout a population. There would thus be a gradual change in the character of the species, as it accumulated changes that caused it to become better adapted to its environment. In other words, the species would evolve.

Natural selection is not a theory of chance. It preserves favorable adaptations and weeds out unfavorable traits. It is the source of all the design that Paley observed in the natural world. For example, natural selection has no trouble explaining the evolution of the eye. If it is advantageous to have better sight, then gradual changes in the structure of the eye will take place.

The eye is believed to have evolved not once, but a number of different times, and, as Darwin noted, all of the stages in the evolution of the eye can be seen in animals that are living today. For example, some single-celled animals have small light-sensitive spots on their bodies. These give them the ability to move toward, or away from, a source of light. Some many-celled animals, such as certain flatworms and shellfish, have light-sensitive cells arranged in a little cup, called an eye-spot. In other animals, such as the swimming shellfish nautilus the eye-spot is deeper, and the opening that allows light to pass into it is narrow, forming a primitive kind of lensless eye that functions something like a pinhole camera.

Once this kind of "eye" has evolved, the evolutionary steps that lead to eyes like our own are not particularly complex. The light-sensitive cup becomes covered with a translucent or transparent layer of skin. In this case the eye cavity becomes filled with cellular fluid rather than air or water. Then part of the region that contains the cellular fluid evolves into a lens. Finally more complex eyes evolve. All of these stages, and some intermediate ones, are observed in various different species of mollusks, while the octopus and the squid have highly developed eyes that are similar to our own.

It is estimated that the evolution of a complex adaptation like a fully functioning eye requires about two thousand evolutionary steps over something on the order of 400,000 generations. In a species with a generation time of one year, a complex eye could evolve in less than half a million years. It should come as no surprise that the eye has evolved independently in a number of different evolutionary lines.

When Darwin formulated his theory, he was hampered by the fact that the mechanisms of inheritance were unknown. Today, this is no longer a problem. We know that the traits seen in living species are expressions of the genes they possess. And we know that random mutations of these genes are the source of the variation Darwin observed. This, incidentally, is the only point at which chance enters into evolution. Once mutations appear, natural selection goes to work preserving the favorable mutations, while weeding out the ones that are harmful. Most mutations are harmful or neutral (neutral in the sense that they neither increase nor decrease the adaptation of an individual organism). However, the harmful mutations are quickly eliminated while the favorable ones remain.

After Darwin, there existed a scientific explanation for the design that was observed in nature and the argument from design lost its force. Darwin's theory of evolution tells us that, if there was a creator, then it is not possible to infer his existence from the characteristics of the numerous different organisms that inhabited the earth. As Richard Dawkins puts it:

> The great beauty of Darwin's theory of evolution is that it explains how complex, difficult to understand things could have arisen step by plausible step, from simple, easy to understand beginnings. . . . Our scientific Darwinian explanations can carry us through a series of well understood gradual steps to all the spectacular beauty and complexity of life.

As we have seen, Dawkins goes further, taking the extreme position that modern scientific knowledge implies that God does not exist. In his 1992 debate with the Archbishop of York, Dawkins made the claim:

> The alternative hypothesis, that it [the universe] was all started by a supernatural creator, is not only superfluous, it is also highly improbable. It falls foul of the very argument that was originally put forward in its favour. This is because any God worthy of the name must have been a being of colossal intelligence, a supermind, an entity of extremely low probability—a very improbable being indeed.

Naturally, not everyone finds Dawkins's argument convincing. The British physicist and Anglican priest and theologian John Polkinghorne objects to this conclusion, saying, "Rather it represents his [Dawkins's] metaphysical judgment on the significance of the scientific story that is presented to us." In *Belief in God in an Age of Science* Polkinghorne argues that the fact we have moral intuitions—that we know what is right and wrong—implies that there is a supreme Source of Value: God.

I won't discuss Polkinghorne's arguments about moral values in detail. After all, the subject of this chapter is what science can or cannot tell us about these matters. Yet it probably won't do any harm to briefly make note of the views of a scientist who occupies the middle ground between the views of Polkinghorne and Dawkins, Harvard paleontologist Stephen Jay Gould, who describes himself as an agnostic. He sees science and religion as "non-overlapping magisteria" that can easily coexist because they deal with different areas of human experience. The magisterium of science covers the empirical realm, he says in *Rocks of Ages*, while religion deals with questions of ultimate meaning and moral value. "To cite the old clichés," he says, "science gets the age of rocks and religion the rock of ages; science studies how the heavens go, religion how to go to heaven."

There is apparently a lot of disagreement among scientists about the existence of God. Some, like Dawkins, are atheists. And some, like Polkinghorne, are devout questioners. Others, like Gould, take the position that science and religion have nothing to do with each other. However, there is one thing on which most agree: Darwin's theory of evolution effectively destroys the arguments made by such proponents of the argument from design as Paley. Nature exhibits no evidence of intelligent design. There may or may not be a God who created the universe. But if there is, it is clear he did not individually design the organisms that make up the biological world.

A Designer Universe?

On April 14 to 16, 1999, the American Association for the Advancement of Science, the largest science organization in the United States, held the "Cosmic Questions" conference, at which scientists were invited to discuss three questions:

1. Did the universe have a beginning?
2. Is the universe designed?
3. Are we alone?

Although the scientists who debated these issues more often than not gave an answer of "No" to the second question, the very fact it was asked is a reflection of the fact that, in recent times, the argument from design has gained newfound respectability. At the beginning of the twentieth century, the argument was generally thought to have been discredited. But today it is again being put forward quite seriously as a quasi-scientific argument by religiously inclined philosophers and scientists.

Oxford philosopher Richard Swinburne sees evidence of design in the beauty and orderliness of the universe. Our universe could very well have been chaotic, he maintains, if a designer had not created laws of nature that maintained order, always operating in the same way. Furthermore, Swinburne says, Darwin did not really refute the idea that nature shows evidence of design. In his book *Is There a God?* he says:

> The [Darwinian] explanation of the existence of complex organisms is surely a correct explanation, but it is not an ultimate explanation of that fact. For an ultimate explanation, we need an explanation of why those laws rather than any other ones operated. The laws of evolution are no doubt consequences of the laws of chemistry governing the organic matter of which animals are made. And the laws of chemistry hold because the fundamental laws of physics hold. But why just those fundamental laws of physics rather than any others? . . . The materialist says that there is no explanation. The theist claims that God has a reason for bringing about those laws because those laws have the consequence that eventually animals and humans evolve.

Swinburne also presents a cosmological argument:

> The matter-energy at the time of the Big Bang had to have a certain density and a certain velocity of recession to bring forth life . . . Increase or decrease in these respects by one part in a million would have had the effect that the universe was not life evolving. For example, if the Big Bang has caused the chunks of matter-energy to recede from one another a little more quickly, no galaxies, stars, or planets, and no environment suitable for life would have been formed on Earth or anywhere else in the universe. If the recession had been marginally slower, the universe would have collapsed in on itself before life could have been formed.

It so happens that this particular argument was already out of date at the time Swinburne wrote these words. Swinburne is correct when he describes the consequences of an expansion that was a little slower or a little faster than it was. However, there are reasons why the universe expanded at the rate that it did.

For billions of years after the universe began expanding, the retarding force of gravity was constantly acting to slow the expansion down. If the density of the universe had been higher, the expansion would have slowed down quickly and the universe would quickly have collapsed into a big crunch. If the density of the universe had been lower, gravity would not have retarded the expansion enough, and the hydrogen and helium gas that filled the universe would have dispersed before it could coalesce into galaxies and stars. However, most cosmologists believe that a tiny fraction of a second after the big bang an inflationary expansion took place. If there was such an expansion, it would have had the effect of fine-tuning the universe so that matter would have just the right density.* In effect, an inflationary expansion pushes a universe to the borderline between one that expands too rapidly and one that quickly collapses.

This doesn't really refute the idea on which Swinburne's argument is based. After all, one can always go back a step further and say that God set the laws of nature so that there would be an inflationary expansion, and so that it would operate in this way. Swinburne did put his finger on a striking fact about the universe in which we live. Our universe does seem to be fine-tuned for the production of life. We seem to live in an improbable universe in the sense that if the laws of physics were just a tiny bit different in any of numerous different ways, then life could never have evolved. In the words of British astronomer Fred Hoyle, it looks almost as if someone was "monkeying" with the laws of physics.

Cosmic Coincidences

The existence of life in our universe seems to depend upon a number of improbable coincidences. If the laws of physics or constants of nature (for example, the speed of light and the strength of the force of gravity)

---

*This is explained in more detail in my book *Cosmic Questions* (John Wiley & Sons, 1993).

were just slightly different, then living creatures could never have evolved. Suppose gravity were just a little weaker than it is. If it were, then galaxies and stars would never have been created. It is gravity, after all, that caused clouds of primordial hydrogen and helium gas to condense into stars.

On the other hand, if gravity were just slightly stronger, the results would be equally catastrophic, at least from our point of view. Under such circumstances the universe might collapse into a big crunch before life had a chance to develop. Or, if for some reason it survived for billions of years as our universe has, it might contain little but black holes. If gravity were stronger then stars would be compressed by gravity to a greater degree. This would cause the pressures and temperatures in their interiors to rise. As a result, the nuclear reactions that cause stars to produce heat and light would proceed more quickly, and stars would have shorter lives, perhaps too short for evolution to produce anything but microorganisms on the planets that surrounded them. In such a case, stars would also be more likely to collapse into black holes. Black holes are produced in supernova explosions, which would exterminate any life on planets that happened to revolve around an exploding star.

Similarly, if the electromagnetic force, which binds molecules together, was too weak, solids and liquids could not exist, and the universe would contain nothing but gas and stars (which are condensed gas). If the electromagnetic force was much stronger than it is, protons would repel one another too strongly and complex atomic nuclei could not form. There seems to be little chance that life would evolve in a universe in which such common elements as carbon and oxygen were not created.

You may recall that the strong nuclear force is the force that binds neutrons and protons together in nuclei. In our universe, this force is just barely strong enough to bind a proton and a neutron together to make deuterium. But it is not powerful enough to create hypothetical particles called *di-protons*, which would consist of two protons bound together. Such particles cannot form because the strong force cannot quite overcome the mutual repulsion of two positively charged protons. But if the strong force were just a few percentage points stronger, then di-protons could be created, and the results would be catastrophic. If di-protons existed, then stars would not burn in the slow and steady manner that they do in our universe. In fact, there would be no such things as stars at all. Concentrations of primordial gas would produce nuclear reactions that progressed so rapidly that immense thermonuclear explosions would take place before stars could even form. In such a universe, hydrogen nuclei would react with one another so readily that the universe would be nearly 100 percent helium today.

If the strong force were just slightly weaker, then no life could exist. If the strong force were just 5 percent weaker than it is, it could not bind a neutron and a proton together and deuterium nuclei would not exist. But the formation of deuterium is one of the steps by which nuclear reactions convert hydrogen into helium, and a universe with a strong force that was too weak would not contain any stars. Gravity would still cause concentrations of primordial gas to contract, but the nuclear reactions that cause our stars to shine would never begin. At best, such a universe would contain some concentrations of gas heated by gravitational contraction. Such bodies would emit small quantities of heat radiation, but for the most part such a universe would be cold and dark.

There is the possibility that some form of life very much unlike us might evolve in universes that were very different from ours. If such a thing is possible, it would be extremely unlikely in a universe that was dark and cold. Life, after all, depends upon the flow of energy. Life would never have evolved on Earth if there were not a constant flow of energy from the sun. A cold, dark universe does not look like a suitable habitat for any form of life, no matter how exotic it might be.

A universe in which the proton and neutron had slightly different masses would be even worse. In our universe the neutron is about 0.1 percent heavier than the proton. As a result, a neutron that is not bound in nuclei will spontaneously decay into a proton, an electron, and a neutrino. This does not happen because the neutron is a composite particle made up of three other particles. The proton, electron, and neutrino are created in the decay process. One particle can decay into three particles because there is extra mass available.

If the proton were the heavier particle, the neutron would be perfectly stable, and protons would decay into neutrons, positrons, and neutrinos. In such a universe, space would be filled with neutrons and little else. The positrons that were created in the proton decays would undergo mutual annihilation with any electrons they encountered. With only neutrons available, the complex structures on which life seems to depend could not be created.

The Balance of the Forces and the Dimensionality of Space

As we have seen, the strength of the strong force must be finely tuned if a universe is to be suitable for the evolution of life. Similar statements can be made about the weak electromagnetic and gravitational forces. However, the balance of the forces—the ratios of one force to another—are also important. I have already given one example of this above. If a universe is to contain life, the electrical repulsion between electrons must not be so strong that it overcomes the effects of the strong force. If it is, nuclei with more than one proton will not be created, and the elements on which life as we know it depends will not exist. Observe that here it is not the absolute strength of the electrical repulsion that is important; what matters is the ratio of its strength to that of the strong force.

The ratio of the weak and gravitational forces is significant also. Certain nuclear reactions that took place in the early universe, in the first few minutes after the big bang, depend upon the strength of the weak force and also upon the rate of expansion of the universe at that time. The creation of deuterium and helium nuclei depends on conditions in the early universe. Altering the gravitational force would change the expansion rate, and thus the rate at which these reactions took place.

If the ratio and the weak and gravitational forces was slightly different from what it is, then the universe would have emerged from the big bang either as 100 percent hydrogen or as 100 percent helium. Now, there does not seem to be any special reason why a universe that was wholly hydrogen at first could not support life. More complex elements could be created in the nuclear reactions that take place in the interiors of stars. These elements could then be ejected into space in supernova explosions, just as they are in our universe (all of the elements heavier than helium were created in this way). On the other hand, it is hard to see how a universe that was originally 100 percent helium could support life as we know it. For one thing, there would be no such thing as water. Water molecules are made of one oxygen and two hydrogen atoms, and hydrogen would not exist in such a universe.

Yet another characteristic of our universe that is crucial to its ability to support life is the dimensionality of space. It is not likely that life would exist if there were fewer or more than three macroscopic dimensions. For example, it is difficult to see how life could exist in a two-dimensional universe; an animal could not possess a digestive tract that ran from one end of it to the other, this would cut it into two pieces. Furthermore, it can be shown mathematically that stable planetary orbits are only possible if the number of dimensions is exactly three. If there were more or fewer dimensions, then planets—if planets could be created in the first place—would either spiral into the sun or drift off into space.

As we have seen, superstring theory *may* suggest a reason why space is three-dimensional. But the calculation that leads to this conclusion is somewhat speculative. It may well be that spaces of other dimensionality are possible. If universes of such dimensionality exist, it is likely that they are either lifeless or contain life of some exotic kind (at least it would seem exotic if viewed with our eyes).

The Anthropic Principle

How does it happen that we live in a universe that is 13 or 14 billion years old? At first this sounds like a senseless question. The universe has to be some age, after all. But upon reflection, we find that the question is more meaningful than it might have seemed to be at first. If the universe was too young or too old, there would be no conscious observers. The universe was originally composed of little but hydrogen and helium gas. All of the elements heavier than helium were created by the nuclear reactions that take place in the interiors of stars and then spread throughout space in supernova explosions. Thus a universe only a few billion years old could not possibly contain life. Entire generations of stars have to be born and die before there are any materials from which planets like Earth can be made. This takes time. So does the evolution of conscious observers. It is believed that the evolution of life on Earth began some 3.5 to 3.8 billion years ago. Billions of years passed before organisms of any great complexity evolved.

However, it is not likely that there would be any conscious observers in a universe that was too old, either. Stars eventually burn out. It is true that new stars are constantly being created, but the supply of interstellar gas from which stars are made will eventually run out. It appears that hundreds of billions of years from now, the universe will be a cold, dark place incapable of supporting life. Thus an age of 13 or 14 billion years falls within a relatively narrow window in which conscious life if possible.

The idea that the universe must have certain properties if conscious observers are to exist is called the anthropic principle. Scientists do not agree about its significance. Some consider it to be a useful concept, while others maintain that it tells us nothing new. The American physicist Heinz Pagels was a harsh critic of the idea, calling it a kind of "cosmic narcissism." According to Pagels, the anthropic principle was not very scientific. However, other physicists have seen great significance in the idea.

The anthropic principle can be expressed in two different ways, either in a "weak" or a "strong" form. The British physicist Brandon Carter has stated the weak anthropic principle as follows: "What we can expect to observe must be restricted by the conditions necessary for our presence as observers." In other words, if the universe did not have certain properties, we could not be here to see it.

If the weak anthropic principle sounds somewhat philosophical, then the strong form is positively metaphysical. Carter expresses the strong anthropic principle as follows: "The universe must be such as to admit the creation of observers within it at some stage." In other words, the universe *must* be of such a character as to provide an environment in which conscious life can evolve.

There is no known way that the strong anthropic principle could be empirically verified. So why do some scientists talk about it at all? I suspect they do because it provides a framework for discussing the fact that the universe has so many different properties that are essential for the evolution of conscious observers. Talking of anthropic principles is a way of summing up all that has been observed about the various different ways in which the universe seems to have been fine-tuned.*

It would be perfectly possible to discuss the fine tuning of the universe without bringing in anthropic principles at all. Scientists like Pagels preferred such an approach. Anthropic principles are discussed fairly frequently these days, however, and it doesn't hurt to have some familiarity with the terminology used. If you are not inclined to bother with this, little will be lost if you skip ahead to the next section.

In their book *The Anthropic Cosmological Principle* scientists John D. Barrow and Frank J. Tipler suggest that if the strong anthropic principle is true, there are three different ways of interpreting it. The first possible interpretation is that the universe was deliberately designed to be hospitable to life, According to the two authors, this view can neither be proved nor disproved because it is religious, not scientific, in nature. The second possibility is that a universe cannot come into being if it is incapable of producing conscious beings that can observe it. Barrow and Tipler attempt to give this somewhat mystical-sounding view a scientific foundation by relating it to certain interpretations of quantum mechanics. This idea is not as fashionable as it was when *The Anthropic Cosmological Principle* was written, and I will refer the interested reader to Barrow and Tipler's book. I personally find speculation about observer-created universes to be metaphysical in the extreme, and I am not sure that relating it to ideas in quantum mechanics makes it any less so. Thus I see little to be gained by pursing the idea here.

The third possibility cited by Barrow and Tipler is somewhat more straightforward. They suggest that if there is a very large number, perhaps an infinite number of different universes, it would not be surprising if conscious beings existed in some of them. According to this

---

*In recent years, variations on the anthropic principle have proliferated. For example, there is the weak anthropic principle, or WAP; SAP, the strong anthropic principle; PAP, the participatory anthropic principle; and FAP, the final anthropic principle. Author Martin Gardner has responded to this profusion by adding CRAP, the completely ridiculous anthropic principle. But since only WAP and SAP are relevant to this chapter, I'll forgo discussion of the others.

interpretation, most universes are indeed lifeless. Our universe has the properties it does because otherwise we would not be here to see it.

A Designed Universe or an Infinity of Universes?

According to Richard Swinburne, the fact that the laws of physics seemed to be designed to allow the evolution of conscious life implies the existence of a creator. He discounts the possibility that there may be other universes with different physical laws, saying, "Every object of which we know is an observable component of our universe, or postulated to explain such objects. To postulate a trillion trillion other universes, rather than one God to explain the orderliness of our universe, seems the height of irrationality."

This is not the kind of argument that often appeals to scientists, whether they are theists or not. Most scientists do not like to discuss religious ideas in scientific contexts. Like Stephen Jay Gould, they generally feel that domains of science and religion are mutually exclusive. And of course, there is a third point of view. Scientists who are atheists are often quite critical of the idea that the universe shows evidence of design. For example, in a talk at the Cosmic Questions conference, Nobel Prize–winning physicist Steven Weinberg, an atheist, described the idea that the laws and constants of nature had been fine-tuned to allow for the evolution of conscious life as "little more than mystical mumbo jumbo." "It seems to me," he said,

> that physics is in a better position to give us a partly satisfying explanation of the world than religion ever can be, because although physicists won't be able to explain why the laws of nature are what they are and not something completely different, at least we may be able to explain why they are not slightly different. For example, no one has been able to think of a logically consistent alternative to quantum mechanics that is only slightly different. Once you start trying to make small changes in quantum mechanics, you get into theories with negative probabilities and other logical absurdities.

In other words, the apparent fine-tuning of the universe was often simply the result of logical consistency.

Weinberg also cited that there might be other universes, or regions of our universe in which the laws of physics were different. "According to the 'chaotic inflation' theories of Andre Linde and others," he said, "the expanding cloud of billions of galaxies that we call the big bang may be just one fragment of a much larger universe in which big bangs go off all the time, each one with different values for the fundamental constants." He went on, "In any such picture . . . there would be no difficulty in understanding why these constants take values favorable to intelligent life."

To me Weinberg's arguments are convincing. There may or may not have been a creator. However, if there was, I suspect that his existence cannot be inferred from the laws of physics. Swinburne argues that it is the "height of irrationality" to postulate the existence of other universes, yet it can equally well be argued that it is the height of irrationality not to do this. There is no reason to think that the big bang that produced the universe we live in was the only one that ever took place. The idea that there is only one universe requires additional assumptions. The idea that there have been numerous big bangs or that there exist numerous other universes does not. It seems to me that the burden of proof is on proponents of the argument of design, such as Swinburne, not on those who express skepticism about it.

The Possible Implications of Superstring Theory

As I pointed out in a chapter 5, one of the motivations for the intense interest in superstring theory is that many physicists hope that such a theory will tell us why the known laws of physics and the constants of nature are as they are. A successful superstring or M theory also might tell us in precisely which ways these laws and constants could vary from universe to universe.

No one knows what will be found in the future. Consequently, arguments about the implications of the precise form of the laws of physics are still on somewhat shaky ground. It is even possible that we might eventually discover that the only laws of physics that are possible are ones that produce universes hospitable to life. I won't venture to speculate what the consequences of such a hypothetical discovery might be. Like many physicists, I tend to doubt that even this would be evidence of design. I suspect it is more likely it will be found that the laws of nature can vary in a number of different ways. Naturally, it is not possible to be sure about this.

If I were to attempt to make predictions about the future, there is only one I would put forward with any confidence. I suspect that arguments about evidence of design will be continue to be made in the future, and that these arguments will be based on some surprising future discoveries about the nature of physical reality.

9 | *What Is the Purpose of It All?*

In *The Demon-Haunted World*, Carl Sagan speaks of finding "spiritual" values in science:

> In its encounter with Nature, science invariably elicits a sense of reverence and awe. The very act of understanding is a celebration of joining, merging, even if on a very modest scale, with the magnificence of the Cosmos. And the cumulative worldwide buildup of knowledge over time converts science into something only a little short of a translational, transgenerational meta-mind.
>
> "Spirit" comes from the Latin word "to breathe." What we breathe is air, which is certainly matter, however thin. Despite usage to the contrary, there is no necessary implication in the word "spiritual" that we are talking of anything other than matter (including the matter of which the brain is made), or anything outside the realm of science. On occasion, I will feel free to use the word. Science is not only compatible

with spirituality; it is a profound source of spirituality. When we recognize our place in an immensity of light-years and in the passage of ages, when we grasp the intricacy, beauty and subtlety of life, then that soaring feeling, that sense of elation and humility combined, is surely spiritual. So are our emotions in the presence of great art or music or literature, or of acts of exemplary selfless courage such as those of Mohandas Ghandi or Martin Luther King, Jr. The notion that science and spirituality are somehow mutually exclusive does a disservice to both.

Steven Weinberg has quite a different opinion, one he expresses in a few succinct words. In *The First Three Minutes*, Weinberg says "The more the universe seems comprehensible, the more it also seems meaningless."

Is science something that can inject values into our lives? Or does it tell us that we live in a vast, often hostile and meaningless cosmos? Certainly scientists can experience elation upon making new discoveries and upon contemplating our growing understanding of the workings of nature. But what about those who are not scientists? I suspect that they experience very little of the "spirituality" of which Sagan speaks. For that matter, I can't help but think that many of the preoccupations of our time are symptoms of a profound distrust of science and of the scientific worldview.

If this distrust were not prevalent, would there be such great interest in such things as astrology, the predictions of psychics, poltergeists, remote viewing, and pseudoscientific ideas about UFOs and crop circles? We live in what is supposedly the most scientific of all ages. Yet dubious and pseudoscientific ideas evoke greater interest than accounts of scientific discoveries. At the beginning of the twentieth century there was not one American newspaper that ran a horoscope column. Now there is hardly a newspaper without one, and there are a great many more professional astrologers than there are astronomers. Distrust of medical science has become so great that droves of people seek out alternative "holistic" remedies. Some of these are truly bizarre. For example, homeopathic preparations are subjected to so many successive dilutions in distilled water that in many cases not a single molecule of the supposedly therapeutic substance remains.

In Victorian times, English working-class people flocked to lectures given by noted scientists. One has to be a guru, a psychic, or a faith healer to attract that kind of audience today. We see television programs that tell us that the moon landings were a hoax and that human beings lived at the same time as dinosaurs. In Kansas, creationists succeeded in setting up educational standards that would have removed evolution, radioactive dating, and the big bang theory from the school curriculum. To be sure, this decision was overturned when the people of Kansas elected a new board of education. However, it did not prevent similar assaults on the presentation of science in schools to be made in other states. All of the evidence seems to indicate that not only do members of the lay public fail to see "spiritual values" in science, but they find much about the scientific worldview profoundly alienating.

It is not my intention to debunk pseudoscientific ideas. Other books have done that, often quite well. Nor do I want to attempt a sociological analysis of the common attitudes toward science in our time. What I propose to do instead is to take a close look at the worldview of modern science and try to see if it really does make the universe seem as meaningless as Weinberg claims, or whether it might at least have the potential to be a source of "spiritual" values, as Sagan thought.

Cosmic Evolution

Scientists do not know precisely how our universe came into existence. There are plausible arguments that suggest it might be the result of a chance event, such as a quantum fluctuation. However, the evolution of the universe from a time of one second to the present is well understood, and there are reasons for thinking we have a correct theoretical understanding of the evolution of the universe at even earlier times.

After the big bang fireball had died out, the universe consisted of little but clouds of hydrogen and helium gas. At this point, the universe was expanding rapidly. Gravity, which acted as a braking force, kept the expansion from being too rapid. If there had been less gravitational retardation, the gas that filled the universe would have become too dispersed to allow the creation of galaxies and stars.

Gravity acted not only on the expansion of the universe, it also acted on the gas clouds themselves. The density of matter was not uniform throughout the universe. Gravity acted to condense the gas in denser regions, gradually causing it to become more compressed. This caused some of the gas clouds to become even denser, and the work of gravity was speeded up. When a large, massive cloud of gas is even slightly compressed, individual gas atoms are closer together, and the gravitational attraction becomes greater. The gas clouds of which I am speaking were incredibly more massive than any that we observe on Earth. Many of them eventually condensed into galaxies containing many billions of stars. For example, there are about 100 billion stars in a typical spiral galaxy, such as our own Milky Way.

Scientists do not understand all the details of galaxy formation. The gravitational effects of dark matter certainly played a role. But physicists and astronomers do not yet know what all of the dark matter is or how it was distributed throughout the universe at the time of galaxy formation. However, it is likely that after gravity condensed galaxy-size clouds of gas, that this gas began to fragment into smaller gas clouds that would eventually become stars.

Star formation began perhaps a billion years after the big bang. It isn't possible to pinpoint the time with any great accuracy. Scientists will have to know more abut galaxy formation before they can do that. However, when the process of compression began to take place a second time, something new began to happen. The contracting gas clouds that were to become stars heated up. Increasing pressure causes the temperature of gases to rise. You have probably noticed the opposite effect, the cooling of a gas as it expands; it is the expansion of the gas inside an aerosol can that makes the can feel cold when some of the gas is released.

The contracting gas clouds were not all the same size. The more massive ones became hotter than the others. When there was more mass, the forces that caused gravitational contraction were greater, and the internal pressure was greater. Greater pressure produced more heat. In many of the clouds, the temperatures became high enough that nuclear reactions could begin. In others—those with masses less than about 8 percent of the mass of our sun—this couldn't happen; temperatures did not become elevated enough. These bodies became *brown dwarfs*. They radiated heat and some light at first, but nowhere near as much heat and light as a true star. The planet Jupiter may be one such "failed star." Jupiter's interior is still hot enough that it radiates away more heat than it receives from the sun. The planet is mainly composed of hydrogen and helium in proportions similar to those observed in the sun. If Jupiter had been a bit larger, it might also have become a star.

No planets like Earth orbited this first generation of stars. Elements such as oxygen, nitrogen, carbon, silicon, and iron that are so common on Earth did not yet exist. All of these elements were manufactured in the cores of massive stars, and then spread through space in supernova explosions.

The energy produced by stars comes from nuclear reactions, which convert hydrogen into helium. Or at least it does initially. The helium fuel eventually becomes exhausted. In average-size stars like our sun, this takes a long time. The sun has been shining for 5 billion years, and another 5 billion years will pass before it begins to die. When the sun's supply of hydrogen begins to run out, the pressure caused by the nuclear reactions in its core will decrease. This will allow gravity to compress the matter of which the sun is composed still further. When this happens, the temperature of the sun's core will rise. This is analogous to the process that causes a star to become hot in its early stages of formation. In both cases, compression produces heat.

When the temperature in the sun's core becomes high enough, a different kind of reaction will begin. Pairs of helium nuclei will fuse to produce nuclei of the metal beryllium. Now, it so happens that the beryllium nuclei that are formed in this manner are not stable. If nothing else happened, they would break apart into helium again, and the net production of energy would be zero. However, if a beryllium nucleus captures another helium nucleus before this happens, carbon will be created. And carbon nuclei created from three helium nuclei are stable.

After the sun converts all of its helium into carbon, its nuclear fires will die out, and it will gradually become a *white dwarf*, a burned-out star that continues to glow only because of its residual heat.* However, temperatures in the cores of stars more massive than the sun rise again when the helium fuel is exhausted, and nuclear reactions similar to the one I have just described will produce heavier nuclei, such as those of oxygen, silicon, and iron. Iron nuclei are the most stable ones of all. Energy is not released when heavier elements are created; on the contrary, energy is required to make them. This explains why such heavy metals as gold and uranium are relatively rare compared to iron.

The higher temperatures that exist in the cores of massive stars cause them to experience more violent deaths than that of our sun. When the nuclear reactions that take place within a massive star finally cease, the star's core collapses, producing a supernova explosion that sends the star's outer layers flying off into space. The explosion is so violent that, for a brief period, a supernova shines as brightly as millions or billions of stars.

All stars shed a considerable amount of matter before they reach the ends of their lives. A star that is initially as much as eight times heavier than our sun may eject enough matter that it will die in a manner similar to that of our sun. But stars heavier than this must eventually explode into supernovas. Massive stars have lifetimes that are much less than our sun; their nuclear fuel is consumed more rapidly. Such stars typically have life spans of hundreds of millions of years, much less than the 10 billion years that is typical of stars the size of the sun. Thus the heavier elements very likely began to be spread through space while the universe was still relatively young, perhaps less than 2 billion years old.

The Creation of Earth

These heavier elements were incorporated into second- and later-generation stars and the planets that formed around them. Our sun is one such star. Our solar system began to condense from a cloud of dust

---

*The sun will actually go through a red giant phase before it becomes a white dwarf. I am omitting some of the details to go on more quickly to a discussion of the creation of elements other than helium and carbon.

and gas about 5 billion years ago. The gas was mostly hydrogen and helium and the dust grains consisted of elements that had been "cooked" in the cores of stars. The initial condensation of this cloud might have been triggered by the shock wave from a nearby supernova explosion. Stars are often created this way today.

As the cloud collapsed, a dense, slowly rotating core formed. This core was to become the sun. It was surrounded by a disk of dust and gas that was spinning more rapidly. The centrifugal force created by this motion prevented it from falling into the sun. Like the gas clouds that condensed into galaxies and stars, the disk exhibited fluctuations in density. Regions that were denser than average exerted gravitational forces on material in their surroundings, and much of the dust came together into small clumps. Calculations indicate that these clumps, which are called *planetesimals*, were about the size of asteroids. At this time no planets yet existed.

Gravity attracted more dust grains, as well as gas, to the largest planetesimals, causing them to become more massive yet. Planets such as Jupiter and Saturn, which are called gas giants because they are composed primarily of gas, have rocky cores. The cores make up only a small part of their total volumes. The inner planets, Mercury, Venus, Earth, and Mars undoubtedly attracted some gas during the early stages of their formation. However, these planets are not massive enough to retain much hydrogen or helium—light gases that easily "boil off" into space. Today there is no hydrogen in Earth's atmosphere, and so little helium that helium was first discovered on the sun rather than on Earth.

Earth was created approximately 4.6 billion years ago. At this time, planets still experienced numerous collisions with asteroids. Studies of the moon, Mercury, and Mars indicate that impact craters were created then at a rate about a thousand times greater than they are today. And every time a planet experienced such a collision its mass increased. During the first 100 million years of the formation of the solar system, the planets increased in size to something close to what they have today.

The collisions that the primordial earth experienced produced so much heat that Earth formed a molten core that was composed mostly of iron. The core is still liquid today. The energy released by the radioactive decay of elements within it keeps it hot. As Earth's iron core formed, lighter elements floated to the surface, creating a rocky mantle. This too remained molten at first; the early Earth had no solid surface.

At first, Earth's atmosphere probably consisted mainly of nitrogen and carbon dioxide. There was little or no oxygen. The oxygen in our atmosphere was created much later by photosynthesizing bacteria. There was no water on the surface at this time, either. Most of the water that is present on Earth today was released from the mantle in volcanic eruptions. This water is believed to have originally come from comets, which are mostly ice. (Comets are sometimes described as "dirty ice balls"; the "dirt" consists of small amounts of various other substances.)

It appears that one could not think of a more inhospitable environment for the creation of life. Not only was Earth extremely hot, it also received a great deal of ultraviolet radiation from the sun. Since there was no oxygen in the atmosphere, there was no ozone layer to screen out the ultraviolet rays (ozone is composed of molecules containing three oxygen atoms; in ordinary oxygen gas, there are two atoms in each molecule).

Yet life evolved, and it evolved quickly. It seems to have appeared almost as soon as conditions became reasonably favorable.

The Origin of Life

Life seems to have evolved with extraordinary rapidity. Impacts with asteroids and meteorites early in Earth's history released so much heat that the surface remained molten for as long as 800 million years. Since Earth has existed for 4.6 billion years, this means that it lacked a solid surface until approximately 3.8 billion years ago. Yet a billion years after Earth was created it was teeming with life. Fossils of organisms resembling blue-green algae have been found in rocks in Africa and Australia that have an age of 3.5 billion years. And chemical traces of life have been found in rocks in Greenland that are 3.85 billion years old. If there really were living organisms at that early a date, they are not necessarily our ancestors. Life might have been wiped out by collisions with large extraterrestrial bodies on one or more occasions, only to begin anew later. However, there is strong evidence that life evolved as soon as the earth crust cooled enough for oceans to form.

Scientists believe that the ingredients for the creation of life were already present at the time that the crust cooled. Many different organic chemicals have been detected in space, including amino acids, the building blocks from which proteins are made. When a cloud of interplanetary gas becomes cool enough, chemical reactions take place, and numerous different substances are produced.

Amino acids could easily have been carried to the earth's surface on falling interplanetary dust particles. They could also have been deposited on the earth when the tails of comets brushed the earth's atmosphere. Most likely both processes took place. And of course amino acids could also have been created on Earth itself. There have been numerous experiments that have shown that this is possible. Scientists are not sure about the exact composition of the atmosphere of the primordial Earth, although most think that carbon dioxide and nitrogen were present in the highest quantities. However, experiments have been performed in various different kinds of simulated atmospheres, and it has been shown that both amino acids and nucleotides—the components of DNA—can be created under a variety of different conditions.

It is not known precisely how, or under what conditions, life began. However, it is clear that some kind of chemical evolution must have taken place. Complex organic chemicals had to be created before life could get a start. It is fairly certain that this did not happen in the midst of the oceans. Amino acids and other organic chemicals were present there only in very small quantities. Furthermore, water tends to break chains of amino acids apart. If any proteins had been formed in the oceans 3.5 billion years ago, they would have quickly disintegrated. However, there were other environments in which life could have originated more readily. Darwin suggested that it might have started in some "warm little pond." This is indeed possible. If the heat of the sun caused water in a tide pool to evaporate, the solution of organic chemicals would have become more concentrated, and reactions between them would have occurred more frequently. Alternatively, life might have begun in volcanic vents in the ocean, or on clay surfaces. If chemicals adhere to a two-dimensional surface, they will meet up with one another more frequently than they will in a three-dimensional space, such as a pond. Finally, life could have begun in rocks and minerals, which could not only have concentrated the organic chemicals but also have protected them from harsh ultraviolet radiation from the sun.

Did life begin with proteins or with RNA (a simpler relative of DNA)? No one knows. The only thing that we can be sure about is that life didn't begin with strands of DNA that arose spontaneously. Today, all living organisms use DNA as a carrier of their genetic code. But DNA is made up of two strands of nucleotides that are intertwined around each other (this is the "double helix" that one so often hears about). If DNA is to reproduce, the two strands must first be split apart by protein enzymes. But the DNA must be split apart before any proteins can be produced. Initially there were no enzymes that could do this. Thus, if some primitive form of DNA was created by chance, it would have been inert.

Both the "life began with proteins" theory and the "RNA first" theory are plausible. Chemical systems that were able to reproduce themselves could have been created from either proteins or RNA. As time went on, these chemical systems could have become gradually more complex, and eventually the first living organisms could have appeared.

Possibly you don't find these ideas satisfying because there is so much guesswork involved. However life *did* originate, and it is possible to make some fairly good guesses about the way this happened. Furthermore, laboratory experiments have shown that chemical evolution is possible. For example, peptides—small protein fragments—have been created that are capable of reproducing themselves and also of mutating into other forms. We can't travel back 3.5 billion years in time to observe what did happen, but with every passing year, scientists have a better understanding of the likely behavior of the first proteins and RNA.

No one knows for sure how likely the evolution of life is, given the right kinds of conditions. It is conceivable that it depends on some rare chance event, and that life exists only in a few other places in the universe. However, most biologists suspect that this is *not* the case and that the evolution of life is practically inevitable if the ingredients for life are present and if conditions are not too hostile. But this supposition will not be confirmed until life is discovered elsewhere—on Mars, for example. This is one of the reasons why there is currently so much interest in looking for life elsewhere in the solar system. If life is discovered elsewhere it will not only confirm what many scientists suspect, it will likely also give us insights into the ways in which life can evolve. If life is discovered elsewhere in the solar system, it is likely to be bacterial life. However, a lot can be learned from bacteria. Deciphering their genetic code would produce large quantities of data and would probably have important implications for our understanding of the evolution of life on Earth.

The History of Life on Earth

Once life evolved, it must have proliferated fairly rapidly. The first organisms to evolve had no natural enemies, and there would have been ample supplies of nutrients in the oceans. Life initially existed only in the oceans, and it was confined to watery environments for more than 2 billion years. Colonies of bacteria began to form on areas of land near the water's edge only about 1.4 billion years ago. Initially, life was not very complex. The cells of multicellular organisms and of such single-celled organisms as amoebas, which are called *eukaryotic* cells, contain complex internal structures, called *organelles*, that are not present in bacterial cells. The DNA of eukaryotic cells is enclosed in a nucleus (lacking in bacteria) and there are many other different kinds of structures as well. Some eleven different organelles in animal cells are not found in bacteria. But there is no need to enumerate them all; a couple of examples should suffice. Both plant and animal cells contain *mitochondria*, which convert energy-rich molecules into forms of energy that can be used by the cell. Plant cells also contain *chloroplasts*, in which photosynthesis is carried out.

As we have seen, there was initially no oxygen in Earth's atmosphere. But then, as photosynthesizing organisms evolved, they began to release oxygen. Significant quantities of this gas were present around 2 billion years ago, and by 1.5 billion years ago, oxygen levels were close to what they are today. The increase in oxygen must have led to an enormous ecological catastrophe. Oxygen, after all, would have been poisonous to organisms adapted to an environment in which mainly nitrogen and carbon dioxide were present in large quantities. However, organisms that were more oxygen-tolerant evolved, and it is from these that most living organisms are descended.

The increase in oxygen made it possible for complex multicellular animals to evolve. A substance called collagen is an important ingredient of connective tissue in such animals. But collagen cannot be manufactured without oxygen. Oxygen is also necessary if such structures such as gills and circulatory systems are to evolve.

The earliest fossils of multicellular animals date from a period around 550 to 670 million years ago. They are found in the Edicarian deposits in Australia and in similar deposits elsewhere in the world. Edicarian fossils are the remains of soft-bodied marine animals such as worms and jellyfish. In many cases the fossils consist of little more than markings left by burrowing worms. Soft-bodied animals rarely fossilize; normally it is only the hard parts, such as bones and shells, that become fossils. Thus the record of early multicellular life is fragmentary.

Then suddenly, around 530 or 535 million years ago, a large number of animals appear in the fossil record. By this time marine organisms (there was not yet any plant or animal life on land) had evolved hard parts, which are more likely to be preserved. So many different kinds of animals came into existence at this time that scientists sometimes speak of the "Cambrian explosion." This event is called Cambrian because it occurred in the geological period of that name. Many of the fossils that are found represent lines that later went extinct, but others are surely the ancestors of organisms living today.

The first terrestrial plants appear during the geological period called the Ordovician, which lasted from 505 million to 438 million years ago. Animals begin to appear on land during the Devonian period, which lasted from 408 million to 360 million years ago. Mammals began to appear a little less than 200 million years ago. Around 250,000 years ago, early forms of *Homo sapiens* evolved. Finally, modern forms, called *Homo sapiens sapiens* appeared abut 100,000 years ago.

Nothing lasts forever, and life is one of the things that doesn't. Earth will not remain habitable for an unlimited length of time. As the sun grows older, it also becomes hotter. At some point, perhaps hundreds of millions of years from now, it will become hot enough that its heat will cause greenhouse gases to be released from rocks. The greenhouse effect will accelerate, and the surface of our planet will become too hot to support life. Earth itself is likely to cease to exist at some point, too. Approximately 5 billion years from now, the sun will expand into a red giant. When it does, it will become so large it will likely engulf the earth. Of course, by this time, Earth will have been lifeless for some time.

It is likely that life will continue to exist elsewhere in the universe for many billions of years after this happens. However, as the universe ages, the stars will gradually burn out and our universe will become a cold, dark place. If any advanced civilizations exist, they may be able to subsist on other sources of energy for some time. However, such sources are not limitless, and life must ultimately come to an end. It is clear that universes like ours progress from an initial state where life could not possibly evolve to one where life will not be able to continue.

No Inevitability in Evolution

Nothing was inevitable about the evolution of human beings, or even of intelligence. Natural selection has no "goals"; it is simply a mechanism that discards what doesn't work and preserves what does. We are not a preordained "end product" of evolution. There are no internal driving forces in the natural world that lead inevitably to the appearance of intelligence. One can easily imagine scenarios in which intelligent life never involved.

There was something about life on the African savannas around 2 million years ago that caused our brains to begin to get large. At later periods, even larger brains began to be advantageous. When mutations occurred that caused brains to get larger, natural selection preserved them. But this has to be attributed at least in part to a chance collision of Earth with an extraterrestrial body.

If a collision with an asteroid had not wiped out the dinosaurs 65 million years ago, mammalian life would not have gained much of a foothold, and it is unlikely that the dinosaurs would have evolved intelligence. While the dinosaurs still ruled the terrestrial world, there were no mammals larger than ferrets; they couldn't compete in the ecological niches the dinosaurs occupied.

Evolution might have taken a different path on many other occasions. The extinction that caused the demise of the dinosaurs was only one of a great number of extinctions Earth has experienced. For example, in the great Permian extinction, which happened about 245 million years ago, 95 or 96 percent of all living species vanished. Those that survived were not better adapted. Natural selection had not adapted any species for the conditions that followed a collision with an extraterrestrial body (it is now thought that the Permian extinction was also caused by such a collision). Survival was at least partly a matter of luck. For example, clams were relatively unaffected by the conditions that produced the extinction, while other mollusks died out. But there doesn't seem to be anything very magical about being a clam, and clams were certainly not more "advanced" than other marine organisms of the period. The most likely reason we are able to eat clams today is that, for some reason or another, the clams were lucky.

If the Permian extinction had not taken place, or if it had happened at a different time, the evolution of life on earth would have been different, and different kinds of plant and animals—some of which might seem very strange to our eyes—might populate the world today. And there is no reason to assume that any of them would be intelligent.

Physicists and other physical scientists who engage in searches for intelligent life elsewhere in the universe often make the assumption that if life evolves then intelligence—and probably a technological civilization—will inevitably occur. But there is no way of being sure that this assumption is correct. After all, intelligence appeared only once on Earth. And of the 50 million or so species now alive, we are the only one that possesses this quality. We can't be sure that intelligence would have inevitably evolved sooner or later. As evolutionary psychologist Steven Pinker points out, the trunk of the elephant is also something that evolved only once. But no one would claim that its evolution was inevitable. It is conceivable that intelligence does not develop in most places where complex animal life evolves. We have no way of knowing for sure. But the conviction that intelligence must arise in numerous different places in the universe is unfounded.

We are not a "higher" form of life than other life forms on our planet. All have been evolving for 3.5 to 3.8 billion years. For that matter, a mammal is not a more advanced form than a fish. Mammals and fish simply followed different evolutionary paths. The very use of the term "higher" implies evolutionary goals, something of which blind natural selection knows nothing.

Evolution and Purpose

A common fallacy equates evolution with progress. It is a legacy of Victorian times, when an almost religious faith in human progress caused the idea to be associated with certain scientific theories, including Darwin's theory of evolution. It was commonly believed that natural selection led inevitably to the creation of "higher" forms of life, culminating in human beings. It was also believed that modern white Europeans were further up the evolutionary scale than people who populated other regions of the world. Today we find that idea racist. However, racist attitudes were almost universally accepted during the Victorian period.

Evolution happens because chance mutations appear, which are then acted upon by natural selection. Nothing guides this process. Natural selection is a blind, unthinking mechanism. The modern view of our place in nature is very different from that which prevailed in pre-Darwinian times. It was then commonly thought that human bodies were the work of a beneficent creator who gave us the form we had and provided equally carefully designed plants and animals for our use. Human beings were at the summit of creation, "a little below the angels."

After Darwin's theory was accepted—and acceptance came quickly, at least among British biologists—scientists were at first disinclined to topple humanity from its high place. Consequently the idea of evolutionary progress was often substituted for divine design. Today, thanks to modern genetics and a better understanding of the ways in which natural selection operates, we know this idea is implausible. The Copernican revolution told us that there was nothing special about our place in the universe. Similarly, modern Darwinian theory tells us that there is nothing special about human beings.

Searching for Meaning

The American philosopher and psychologist William James published *The Principles of Psychology* in 1890, over a century ago. Although much of the psychological theory in the book is hardly up to date, it is a scientific classic that still repays reading. James had numerous insights into human behavior and thinking that still seem fresh when we read his descriptions of them today.

In volume 2 of his book, James discusses human beliefs at great length. "In its inner nature," he says, "belief, or the sense of reality, is a sort of feeling more allied to the emotions than anything else." According to James, human belief is not based on logic; on the contrary, it has an emotional foundation. "We would believe everything if we only could," he goes on and comments that emotionally exciting ideas are the ones most likely to elicit our belief. In other words, we believe what makes us feel good to believe.

When examined objectively, the scientific account of the evolution of the universe and of human intelligence looks like something that would excite no one but a scientist. The universe may have come into existence by chance. Cosmic evolution and biological evolution are both the result of the blind workings of the laws of nature of the universe in which we live. Evolution did not culminate with the appearance of human beings, and it is possible to invent numerous evolutionary scenarios in which *Homo sapiens* never appears. The scientific worldview is a reductionist account in which the things we value most highly have no place. It seems to say nothing about any purpose or meaning in human life, nothing about morality, nothing about spiritual feelings, nothing about human values.

In other words, the scientific worldview provides little to arouse our emotions. Carl Sagan may be able to see grandeur in the scientific vision of the cosmos. However, I think it likely that most laypeople do not. Is it so surprising, then, that many prefer to believe instead in the miraculous? According to a Gallup poll conducted in 1996, 79 percent of all Americans believe in miracles. According to an earlier Gallup poll, conducted in 1990, 52 percent believed in astrology; 33 percent in the lost continent of Atlantis; and 65 percent in Noah's flood. Forty-one percent believed that dinosaurs and humans lived simultaneously; 42 percent thought it was possible to communicate with the dead; and 67 percent believed they had had a psychic experience. Only 22 percent believed that aliens had landed on earth, but if the same poll were conducted today, the figure would undoubtedly be much higher—we have all heard a great deal about alien abductions during the years since 1990.

To the scientist or skeptic, the people who believe in these things may seem naive. However, there is probably more to it than that. If William James's analysis is correct, they must derive some emotional satisfaction from believing in such things. It is fairly obvious that the creationists who attack scientific ideas—not only evolution, but also the physics of radioactive dating and the big bang theory—do so because they have a deep emotional commitment to fundamentalist religion.

Yes, there is grandeur in the scientific worldview, but it takes a better-than-average amount of scientific knowledge to appreciate it. Even then it tells us nothing about "the meaning of it all." A sober appraisal of the scientific picture of the universe must lead us to conclude, as Weinberg does, that the universe really is meaningless.

I don't mean to imply that there is no meaning in life, or that we should discard our notions of morality. The scientific worldview doesn't imply that we must abandon our ethical standards. I think we would all be horrified by someone who tortured cats for fun. We would consider him abnormal and in need of help. And if he tortured children instead, we would be even more horrified. Most of us—I suppose sociopaths have to be excluded—observe certain rules of moral conduct, and most of us do find meaning in our personal lives. I would only maintain that knowing a lot about the universe is not the thing that allows us to do so.

10 | *What Is the Human Condition?*

At first this question seems too broad and philosophical to be subjected to scientific scrutiny. After all, we don't generally ask physicists how to go about living our lives. We don't ask chemists how to find meaning in our worlds or expect mathematicians to tell us what our ethical codes of conduct should be. We don't even ask such questions of people in the social sciences, not unless we happen to be undergoing therapy with a clinical psychologist. And the questions that we might ask a therapist rarely have much to do with what philosophers have thought of as the "human condition." Anthropologists might be able to tell us something about the varieties of human culture, and sociologists might have some special understanding of our own culture. However, we generally don't look to them for insights about what it means to be human. These are more likely to be found in the classics of literature or the writings of philosophers. It would seem that questions about the human condition are among the ones most likely to remain outside the realm of science.

But perhaps it is not as hopeless as all that. Science can tell us something about that thing we call human nature. In recent years scientists working in the relatively new field of evolutionary psychology have been studying the role played by genetic influences in human thought and behavior. By doing so, they have achieved some genuine insights. The research they are conducting cannot help but give us a better understanding of ourselves.

We are generally not aware of the ways in which our genes influence the things that we do. We rarely stop to think about the fact that human beings are hierarchical, for example. Yet we defer to high-status individuals and are very good at noticing displays of high rank such as the types of cars that people drive and their attire. Similarly, we tend to think we are not territorial in the manner that many animals are. Yet we are quick to become annoyed or angry when someone "infringes on our turf." Our homes are our "castles"; they cannot be entered without an invitation (or a search warrant). Doing so is "breaking and entering."

That men should find young women attractive and women should most often be attracted to men who are older than they are seems so natural that we rarely stop to think about the genetic causes of such preferences. We are not surprised to see women barely over the age of eighteen gracing the centerfolds of *Playboy* and *Penthouse*. Nor are we surprised to see the actor Sean Connery remain a sex symbol past the age of seventy.

We expect favors and invitations to be returned and never stop to think that expectations of reciprocity might have some genetic underpinnings. Parents normally care very much who their offspring marry, and are more likely to raise objections if a daughter announces plans to marry someone "unsuitable" than object to the mate choice of a son. Men are usually disturbed by the idea that a spouse might be having sex with someone else. Women tend to be less bothered if there are no signs of strong emotional involvement. "She really didn't mean anything to me," men tell their wives. But women rarely say this to their husbands. They know that their husbands wouldn't care.

During much of the twentieth century, it was customary to believe that much of human nature was learned. The doctrine of cultural relativism, espoused by such cultural anthropologists as the German-American Franz Boas and his student Margaret Mead, told us that we were shaped by our cultures. The proponents of cultural relativism didn't deny that our biological heritage was important. It provided us with sex drives, for example, and caused us to experience such things as hunger and fear. However, it was culture that determined the ways in which our sexuality was expressed, determined what kinds of things we would eat, and molded social systems in many other ways as well. In other words, although infants everywhere were the same, adults behaved the ways they did because their cultures had taught them to. The human mind, the anthropologists believed, was a kind of blank slate upon which the social practices and beliefs of different cultures were imprinted.

Margaret Mead and Samoa

Of all the anthropological studies that seemed to confirm this view, the most famous and most influential was one that Margaret Mead, reported in *Coming of Age in Samoa,* which was published in 1928. According to Mead, Samoan culture lacked many of the characteristics that were common in Western societies. The Samoans had no status hierarchies, Mead said. There was no tension about sex, and adolescent turmoil was unknown. Girls postponed marriage "through as many years of casual lovemaking as possible." The Samoans experienced no feelings of jealousy and cuckolded husbands exhibited no anger. Here was a culture that was clearly unlike Western culture. The implications of this were obvious: human beings were infinitely malleable. It was their socialization by their cultures that made them behave the ways they did. There was no innate, genetically determined human nature.

Mead's book was followed by another, *Sex and Temperament in Three Primitive Societies*, published in 1935. In it Mead described her studies of three primitive societies in New Guinea, known as the Arapesh, the Mundugumor, and the Tchambuli. She found that the Arapesh were a gentle, unassertive, cooperative people. Interactions among the Mundugumor, on the other hand, were characterized by suspicion and hostility. In the third society, that of the Tchambuli, gender roles were practically the reverse of those that were common in Western culture. For example, the women were believed to be more highly sexed than the men. The women were more sexually aggressive, while the men were passive, responsive, and interested in children.

These and other writings by Mead were enormously influential in establishing the idea that cultural influences were all-pervasive. She had apparently found a number of different peoples whose cultural practices and values were much unlike those found in the West. Furthermore, conceptions of human nature varied from place to place. Some of the things that were taken for granted in our culture would have seemed outrageous to a Samoan, or an Arapesh or Tchambuli. It was culture that molded human beliefs and behavior, Mead claimed, and cultural practices varied more widely than anyone had imagined.

However, in 1983 doubt was cast about the accuracy of one of Mead's anthropological studies. That year, anthropologist Derek Freeman published a book titled *Margaret Mead and Samoa.* Freeman pointed out that Mead spent only nine months in Samoa and that she hadn't spoken Samoan when she arrived there. She hadn't lived among the Samoans and had depended on interviews with informants in order to learn abut the culture. According to Freeman, who lived in Samoa for almost six years, Mead's description of Samoan society was misleading and inaccurate. Freeman found that, contrary to what Mead had claimed, virginity was highly valued in Samoan society. The Samoans had a virginity test and practiced ceremonial defloration of girls at their weddings. Freeman interviewed some of the women who had been Mead's informants. One of them told him she had been "just joking" when she told Mead about the nights that the girls spent with boys. Another confirmed this, saying they had invented the stories for fun.

Freeman found that the Samoan society, which Mead had described as free of tension, was actually one in which rape and reprisals by the rape victim's family were common. Mead had said there was no adolescent turmoil, but Freeman and other anthropologists discovered widespread adolescent resentment and delinquency. Mead's findings, Freeman concluded, didn't stand up to scrutiny.

It is possible to express a few reservations about Freeman's work. For example, he studied a different village than Mead did, and did his work some four decades later. Mead's informants were old women when he interviewed them. An elderly woman might tell a high-ranking man a different story from the one she had told a young woman when she was a girl. Nonetheless there is undoubtedly a great deal of truth in Freeman's critique. Other anthropologists have also cast doubt on some of Mead's assessments of other societies. For example, Mead's "gentle" Arapesh were discovered to have been cannibals.

Since Mead's studies were among the founding documents of the doctrine of cultural relativism, the case for the idea that human behavior is infinitely malleable has been considerably weakened. Because Mead and anthropologists like her were so influential, it was assumed for most of the twentieth century that there was no basic human nature, that people in different cultures behaved and thought the ways they did because they assimilated the values and beliefs of their societies. The discovery of flaws in Mead's work opened up the possibility that some aspects of human behavior might be innate, that the ways in which we behave might be influenced by our genes.

Love and Sex in Thirty-Seven Cultures

According to psychologist David Buss, certain patterns in male and female sexual behavior and mate preferences appear in numerous societies in different parts of the world. In 1989 Buss published a study of mate and sexual partner preferences in some thirty-seven different cultures. He and his fifty collaborators interviewed 10,047 individuals on six continents and five islands. The participants came from polygamous societies, such as those in Nigeria and Zambia as well as from cultures that were more monogamous. Scandinavian countries, in which unmarried couples frequently lived together, were included in the study, as well as nations such as Bulgaria and Greece, where this practice is frowned upon. The questionnaires were administered by local residents who spoke the native language, and people living in both rural and urban areas were questioned. Some of the respondents were well educated, while others had little education; their ages ranged from fourteen through seventy.

Buss's study of human mating preferences was the largest ever carried out. He found that certain kinds of preferences were found in all the cultures. In all the societies, men placed a higher value on youth and physical attractiveness of potential mates than women did. The women, on the other hand, were more interested in a man's economic resources. In none of the societies were postmenopausal women considered to be attractive, but men of the same age did not experience difficulty attracting mates. Women preferred men who were of average or greater height, and they rated dependability as a desirable quality more often than men did.

These findings provided evidence for the idea that certain aspects of human mating behavior do not vary from culture to culture. On the contrary, these preferences are innate. Furthermore, if they are innate, they must have been created by natural selection. Those of our ancestors who had these preferences must have had more offspring than those who did not. For example, if men who preferred older women or who had no age preferences had fewer offspring than those who preferred the young women, then in time a preference for young women would become nearly universal. According to Buss, we are genetically influenced to look for certain qualities in potential spouses. These qualities make them seem more attractive.

Evolutionary Psychology

This kind of reasoning is an example of what is called *evolutionary psychology*. Evolutionary psychologists like Buss look for human behavioral traits that do not vary from culture to culture. They believe that doing so allows them to gain a greater understanding of the evolutionary underpinnings of human behavior. Unlike the extreme cultural relativists among the anthropologists, they are not interested in behavior that varies from one society to another, because studying these tells them little or nothing about our evolutionary makeup.

Evolutionary psychology is a relatively new discipline that is an outgrowth of sociobiology, a scientific discipline that was inaugurated in 1975, when entomologist Edward O. Wilson published *Sociobiology*. The subject was defined by Wilson as "the systematic study of all social behavior." According to Wilson, evolutionary ideas could be used to better understand the behavior of all social animals, whether they were bees or ants or human beings. Evolutionary psychology differs from sociobiology in that only human behavior is studied. Evolutionary psychologists look for human behavioral traits that are universal, and try to explain them by showing that they must have been adaptive for our hunting-gathering ancestors on the African savannas hundreds of thousands of years ago.

Human civilization has existed for a relatively short time. Agriculture was invented only about 10,000 years ago, and it is agriculture that created the concentrations of population that made civilization possible. For most of our evolutionary history, our ancestors lived in tribes of hunters and gatherers, and it is that kind of life to which we are adapted. Ten thousand years is too short a time for any significant evolutionary change to take place. Consequently, the life to which natural selection caused us to be adapted is similar to that of the preliterate peoples who still live as hunters and gatherers today.

According to Buss, this makes it possible to understand why humans in all cultures had the kinds of mate preferences that they did. For example, during the time of our ancestors there must have been a strong correlation between physical attractiveness and health; disease and infection with parasites was common. The men who preferred healthy-looking partners would have had more offspring than those who didn't have this preference. Thus certain innate standards of feminine attractiveness would have evolved. The men who were attracted to young women would also have had more offspring. If some were attracted to postmenopausal women, they would not have had any offspring, and whatever genetic factors caused this attraction would not have been passed on. Similarly, those women who were attracted to men who were dependable, willing to commit to a relationship, and able to help provide for offspring would have been more likely to pass along their genes than women who did not have this preference. Finally, women should continue to exhibit such preferences today.

Buss and his colleagues have followed up the large cross-cultural study with more than fifty other studies of human mating and sexual behavior. They have studied the differences between preferences in long-term mates and short-term sexual partners, such behaviors as jealousy and the behavior of people trying either to keep or get rid of a mate. They have studied male-female conflicts and a number of other traits as well. Buss believes that in most of these cases, it is possible to find evolutionary reasons why we behave the ways we do. Natural selection, he says, has made us the kind of people we are, at least in our dealings with the opposite sex.

Criticisms of Evolutionary Psychology

The most vocal critic of evolutionary psychology is Harvard paleontologist Stephen Jay Gould. According to Gould, if human behavior is adaptive, it does not necessarily follow that it has a genetic origin. In humans, he says, adaptation can come about by the nongenetic route of cultural evolution. Since cultural evolution is so much faster than genetic evolution, he points out, its influence should be more important than any "genetic programming."

Gould believes that there are genetic factors that influence human behavior, but that these factors have more to do with potentialities than specific behaviors. For some reason having to do with life on the African savannas, he says, our brains became big. And once they did, they became able to do things that smaller brains could not do. He gives the example of our ability to read and write. These could not be evolved abilities, he says. The invention of writing is too recent to have been influenced by natural selection.

Gould has charged that evolutionary psychologists make up "just so" stories about the human behavioral traits they study. They use "the old strategy of finding an adaptationist narrative (often in the purely speculative or storytelling mode) to account for genetic differences built by natural selection," he says. As an example, he cites an often-used explanation of why we have a sweet tooth. It is frequently asserted that this must have evolved because it was adaptive for our evolutionary ancestors. It would have caused them to prefer ripe fruit over unripe fruit, which has a lower sugar content. But this is "pure guesswork in the cocktail party mode," Gould charges. It is storytelling that is unsupported by any evidence.

Finally, Gould claims that the strategies used by evolutionary biologists are "untestable, and therefore unscientific." Much of it consists of a search for ideas about the manner in which certain traits might have evolved in ancestral environments. "But how can we possibly know in detail what small bands of hunter-gatherers did in Africa two million years ago?" Paleoanthropologists have discovered some bones and tools that our ancestors have left, he points out, but these do not tell scientists anything abut our ancestors' concepts of kinship, about the size and structure of the groups in which they lived, the relative roles of males and females, or about "a hundred other aspects of human life that cannot be traced in fossils."

Pregnancy Sickness

Gould may be right about the character of some speculation in the field of evolutionary psychology. However, his blanket condemnations clearly do not apply to all the work that is done in the field. For example, biologist Margie Profet has done a convincing analysis that explains the origin of pregnancy sickness (commonly called "morning sickness"). Many pregnant women become nauseated after eating certain foods and thereafter avoid them. This has traditionally been explained as a side effect of hormones. However, Profet uncovered evidence that indicates this explanation is not correct, that pregnancy sickness is an evolved behavioral trait.

Profet noted that nausea protects us against some toxins. Food that contains them is vomited up before it can do much harm, and we are likely to develop aversions to the food that made us vomit. Profet therefore wondered whether pregnancy sickness might not have evolved because it protected a developing fetus against certain toxins. Most plants contain toxins of some kind; these toxins protect them against herbivores and against insects. Animals who eat too much of many plants will experience toxic effects, and some other plants are so poisonous that ingestion is likely to cause death. Animals, including human beings, have evolved defenses against the toxins in turn. The liver detoxifies many of the poisons we ingest, for example. However, a fetus in the early stages of development is more vulnerable to poisons ingested in food. Pregnancy sickness looks very much like a defense mechanism against them.

Toxins are ubiquitous. They are present in many of the foods we eat. For the most part these are not exotic foods. On the contrary, toxins are present in apples, bananas, cabbage, celery, cherries, nutmeg, oranges, potatoes, and soybeans, and numerous other foodstuffs. Eating seemingly innocuous foods during a certain stage of pregnancy can have profound effects. For example, it is suspected that toxins in potatoes can cause neural cord defects. Countries with high potato consumption also have high rates of this kind of birth defect.

Much of the research that has been done on this subject has been carried out in animals. For example, it has been shown that toxins in eggplant can cause birth defects in hamsters. It can even be dangerous to drink the milk of an animal that has ingested foods with certain toxins. In one rural California family a litter of goat kids, a baby boy, and a litter of puppies were all born with severe bone abnormalities after the pregnant woman and a pregnant dog drank the milk of the family's pregnant goat after the goat had foraged on the plant lupine.

In order to test her hypothesis, Profet examined hundreds of different studies related to the topics of pregnancy sickness, of toxins in food, and of the effects of these toxins on a developing fetus. She found that toxins easily tolerated by adults can cause birth defects or induce abortion. She noted furthermore that pregnancy sickness begins when the fetus becomes vulnerable to these toxins and ends when its need for nutrients for growth becomes more important than any lingering vulnerability. She found that women experienced changes in their sense of smell during pregnancy, and that these changes promoted the avoidance of toxins. Finally she found that women who had moderate or severe pregnancy sickness had greater pregnancy success rates than people who experienced only mild symptoms.

The evidence looks convincing. Pregnancy sickness is an evolved behavioral trait, which must have evolved because it was adaptive. Among our ancestors, women who did not experience pregnancy sickness would have been more likely to spontaneously abort or to give birth to children with birth defects. They therefore would have passed along fewer of their genes to succeeding generations than women who did become sick during pregnancy. It was natural selection that preserved the genes that are responsible for this trait.

Mental Modules

Two University of California at Santa Barbara scientists, psychologist Leda Cosmides and anthropologist John Tooby, have taken the lead in formulating the theoretical ideas on which evolutionary psychology is based. Cosmides and Tooby (who are married to each other) compare the human brain to a computer that is made up of a number of different modules that perform specialized tasks. The human brain, they say, contains "mental modules" that cause human beings to behave in an adaptive manner. "Our neural circuits," Cosmides and Tooby say, "were designed by natural selection to solve problems that our ancestors faced during our species' evolutionary history."

Cosmides and Tooby believe that we have specialized neural circuits for such things as learning language, recognizing faces, analyzing the shapes of objects, judging distance, selecting mates, seeking foods to remedy nutritional deficiencies, and for doing countless other things. They believe that knowledge about the specific character of these circuits can be gained by studying human behavior. The principles of evolutionary psychology, they claim, can be applied to any psychological topic, "including sex and sexuality, how and why people cooperate, whether people are rational, how babies see the world, conformity, aggression, hearing, eating, hypnosis, schizophrenia, and so on." They say that anyone who is attempting to understand human behavior should ask the following questions:

1. Where in the brain are the relevant circuits and how, physically, do they work?
2. What kind of information is being processed by these circuits?
3. What information-processing programs do these circuits embody?
4. What were these circuits designed to accomplish (in a hunter-gatherer context)?

In other words, if we want to understand why human beings behave the way they do, we should look for specialized brain modules that induce them to do these things and try to understand why such behavior would have been advantageous in a hunter-gatherer society.

If you think this sounds like an ambitious research program, you're probably right. Neurophysiology has not advanced to the point where it would be possible to find the kinds of neural circuits Cosmides and Tooby postulate, circuits that govern specific behaviors, if indeed they exist. More is known about language processing in the human brain than is known about the neural correlates of any other behavior. And there is still a lot about language processing that is not yet understood. Scientists have long known that two brain regions, known as Broca's area and Wernicke's area are involved in language processing. But they don't know how language processing activity is distributed through other parts of the brain. Neurophysiologists have found areas of the cortex that are associated with production of consonants and vowels, and with the naming of animals and tools. However, it will take years, probably decades, before the brain's processing of language is completely understood.

As a result, most research in evolutionary psychology has centered around question 4. Evolutionary psychologists have concentrated on trying to find adaptive explanations for certain characteristic human behaviors, simply assuming that the appropriate neural circuits exist. Nevertheless, some of the work that has been done in the field, such as Buss's study of human mate preferences and Profet's analysis of studies relevant to pregnancy sickness, have uncovered instances of human behaviors that seem to be innate. No one as yet knows what combinations of circuits in the brain cause these behaviors, but they certainly look like behavioral traits that have been molded by natural selection.

Big Kiku Wants You to Have a Tattoo

Suppose you are shown four cards that are marked with the following symbols:

D F 3 7

You are then asked which two cards you must turn over to see if any of them violate the following rule:

> If the letter D is on one side, then there will be a numeral 3 on the other side.

Which two cards would you turn over? If you are like most people, you will give the wrong answer. More than 75 percent given this test answer incorrectly. Even people who have taken courses in mathematical logic do poorly. The correct answer is that one must turn over the card with the D and the one with a 7. If the D card does not have a 3 on the other side, the rule is violated. If the 7 card *does* have a D on the opposite side, the rule is also violated. This experiment is called the Wason selection task after the psychologist Peter Wason, who devised it. It is supposed to be a test for logical thinking, and the results seem to indicate that human beings do not think very logically.

But wait a minute. Suppose you are asked the following question, "If you are a bouncer in a bar, and if only people eighteen or older are allowed to drink beer, which do you have to check: a beer drinker, a Coke drinker, a sixteen-year-old, or an eighteen-year old?" Most people correctly answer that the bouncer must make sure that the beer drinker is eighteen or older, and he must make sure that the sixteen-year-old is not drinking beer. This test is logically the same as Wason's. The only difference is that beer drinking, Coke drinking, and ages have been substituted for letters and numbers. Yet people's success rates are quite different.

It was Leda Cosmides who discovered that people could perform the Wason selection task quite well if it was put in certain kinds of social contexts. She has also found that if the context is somewhat more exotic, people still do well. For example, in an experiment devised by Cosmides and Tooby, the subjects were asked to imagine a powerful chief named Big Kiku on an island in the Pacific. Big Kiku had a habit of demanding that his followers tattoo their faces. One night, four hungry men stumbled into his village and asked to be fed some cassava root. Big Kiku replied that if they tattooed their faces, then they would be fed in the morning. Now suppose that Big Kiku tells you sometime later that the first man got a tattoo, while the second was given nothing to eat. The third man did not tattoo his face, and the fourth received a large cassava root. Which of the men must you inquire about further to see whether or not Big Kiku kept his end of the bargain. About 75 percent of people given this problem correctly answer that they must ask Big Kiku whether the first man (who got the tattoo) was fed, and whether the second man (who went away hungry) got a tattoo. The other two cases are not relevant because Big Kiku would not have broken his promise if he refused to feed the man without a tattoo, or if the fourth man (who received a cassava root) had not been tattooed. Big Kiku could have fed the fourth man at whim without breaking any promises.

Cosmides, Tooby, and other evolutionary psychologists devised other forms of the Wason selection task, and used them to perform various kinds of experiments over a period of eight years. The two scientists say that their results indicate that the human brain contains a "cheater detection" mechanism that has evolved for the purpose of determining whether bargains and social contracts had been adhered to. They claim to have found one of the mental modules that evolutionary psychologists seek to discover.

According to Cosmides and Tooby, the experiments undermined the notion that ideas about social exchange were culturally transmitted. The cheater detection mechanism looked more like something that was innate. People automatically made certain assumptions about social contracts. Thus they found the Wason selection task much easier when it was placed in social contexts that involved the detection of cheating or the making of bargains.

If an evolved cheater detection module exists in the human brain, then it must have evolved because it facilitates human cooperation. There are many circumstances in which participation mutual sharing is advantageous to the individual. Our ancestors on the African savannas would undoubtedly have found it advantageous to offer food to others in times of need if they could reasonably expect that the favor would be returned when they were hungry. This kind of cooperation would only have worked if our ancestors were wary of cheaters.

In order to see why this should be the case, consider a situation in which some individuals are genetically disposed to behave altruistically while others are more selfish. Under such circumstances, the genes that caused altruism would disappear over a period of many generations. Those who accepted the gifts of the altruists would be more likely to survive and reproduce. On the other hand, the altruists would incur costs every time they helped other individuals, and they would be less likely to survive and reproduce than they would have been if they had done nothing. The genes that caused altruism, in other words, would be weeded out by natural selection.

On the other hand if some individuals were genetically disposed only to offer help to others when they had some expectation the favor would be returned, they would have a selective advantage. They would be more likely to be helped when they needed it than would individuals who did not share. The genes that induced such behavior would spread through a population, with the result that most individuals would be concerned about the possibility of cheating.

Most human beings no longer live as hunter-gatherers on the African savannas. However, their behavior seems to be governed by the same principles that might have governed the behavior of our ancestors. When we perform favors, we generally expect to get something in return at a later date. If I invite you to a dinner party, and you accept the invitation, I am likely to feel slightly miffed if I don't receive a return invitation at some point. Similarly, close friends or relatives who give each other Christmas presents or send Christmas cards are likely to feel puzzled or angry if one year, for no apparent reason, the gesture is not returned. If you do a favor for someone you work with, you will probably expect to receive a return favor from that person sometime in the future. You may not be consciously thinking of this when you help your coworker, but you are likely to be puzzled or annoyed if that person refuses to give you some help when you need it.

This all sounds plausible. However, there appears to be little hope of obtaining any solid evidence that would overwhelmingly confirm such ideas anytime soon. We can't travel a million years back in time in order to study cooperative behavior among our ancestors. And neurophysiology has not advanced to the point where it would be possible to find neural circuits in the brain that caused human beings to engage in sharing behavior. Even though Cosmides and Tooby conducted a wide variety of different kinds of experiments with variants of the Wason selection task, it is still possible to maintain that human beings do each other favors and resent cheaters because their cultures teach them to do these things. The evidence that Cosmides and Tooby have uncovered allows them to make a strong case for the idea that resentment of cheaters is innate. However, their evidence is not overwhelming.

Much of evolutionary psychology has this character. There have been some studies, such as Profet's observations on pregnancy sickness, that are hard to dispute. However, the greater part of the work in the field is of such a nature that it is not likely to satisfy critics such as Gould. The findings of the evolutionary psychologists have been suggestive and often quite plausible, but it will be necessary for scientists to learn more about the detailed workings of the brain before they can be said to be absolutely convincing. The pioneers in evolutionary psychology have shown that at least some of our behavior is genetically influenced. But it remains to be seen whether there are as many innate components as they believe.

The Implications of Evolutionary Psychology

It seems clear that at least some of our behavioral traits are genetically influenced. Evolutionary psychologists have not disproved the idea that we engage in certain kinds of behavior because our cultures inculcate certain beliefs and value in us. Indeed, there is ample evidence that these cultural influences exist. If they didn't, then human cultural practices would not vary so widely. On the other hand, some of the ideas developed by evolutionary psychologists seem so compelling that it is necessary to conclude that the behavior of human beings is influenced by their genes to a greater extent than had been previously thought.

It doesn't follow that our behavior is "controlled" by our genes. As thinking animals we have the ability to behave in ways that are contrary to our genetic makeup. My genes tell me that, as a heterosexual male, it would be desirable for me to sire offspring. This is something I haven't done, and since I am past the age of sixty, it is not likely I will. Our genes generally urge us to eat foods that contain a lot of saturated fat. This would undoubtedly have been adaptive for our ancestors, who could not pick up a steak from the supermarket whenever they felt like it but who needed to store fat in their bodies to get themselves through the inevitable lean times. However, nowadays many people realize that it is best to resist these urgings and to maintain a healthier diet. It is in the genetic interest of men to attempt to impregnate as many women as possible. However, few of us do that today. If we did, those men who were married might find themselves suddenly divorced and wind up paying large sums in child support as well. If some man did try to impregnate a lot of women, he would likely be frustrated by the fact that women are not under the control of their genes, either, and are therefore likely to use contraception.

If our genes do not control us, they nevertheless influence our behavior in ways of which we are unaware. For example, the degree of plumpness that is considered attractive may vary from culture to culture. However, according to studies conducted by psychologist Devendra Singh, men consider certain waist-to-hip ratios to be more attractive than others. Women with ratios of 0.70 are considered more attractive than those with ratios of 0.80. The latter are seen as more attractive than women with ratios of 0.90. Singh also measured the waist-to-hip ratios of beauty contest winners and *Playboy* centerfolds over a thirty-year period. Although both centerfolds and beauty contest winners got thinner over this period, their waist-to-hip ratios remained constant at 0.70. Numerous studies show that women with higher ratios have, on the average, more difficulty becoming pregnant, and waist-to-hip ratio can also be an indicator of overall health. Thus it seems likely that men consider certain waist-to-hip ratios sexier than others because their ancestors were more likely to sire offspring with the women who had them. But of course few of them are conscious of this. It is probably safe to say that, when men look at *Penthouse* or *Playboy* centerfolds, having offspring is not the first thing that comes to mind.

If the ideas of the evolutionary psychologists are not totally wrong— and this seems unlikely—then an essential aspect of the human condition is to find ourselves urged to behave in certain ways without knowing why. Our genetic urgings seem so natural to us that we are often not aware they even exist. For example, I like rare steaks, but I rarely stop to think that I probably have genetic programming that causes me to enjoy them. Women are attracted to men who have high status, or who command a lot of resources (one of the preferences confirmed by David Buss's study), but they rarely think of those men in such terms; more often they will simply say that they find certain men sexy. Biologist George Orians has done studies which showed that both children and adults liked landscapes that resembled African savannas. But I would venture to guess that if you asked a woman (or a man) why she enjoyed a certain landscape, she would simply reply that it was "beautiful" and would believe no further explanation was needed.

Yes, there is a basic human nature. Some of our behavioral traits are a result, not of socialization into our cultures but of natural selection. They are products of our evolutionary history. As a number of studies by evolutionary psychologists have shown, some human behavior traits are universal. They do not vary from culture to culture. Possessing these traits is part of what it means to be human.

11 | *What Is Mind?*

I recently had an interesting conversation with a chatterbot named ALICE. A chatterbot, or chat robot, is a computer program that is designed to talk with people over a computer terminal, or over the Internet. "ALICE" really isn't a name, it's an acronym for Artificial Linguistic Computer Entity. This program is the creation of Richard S. Wallace, a robotics researcher and former director of the robotics architecture group at a company called Neuromedia in San Francisco.

While talking to ALICE, I suspended my disbelief and behaved as though I were talking to a real, intelligent person, one that happened to live inside a computer. Our conversation began like this:

>How are you today, Alice?
Ah. Everything is going extremely well.

>Do you talk to a lot of people, Alice?
Yes. A lot more than you might think.

>How long have you been chatting with them?
I am the product of over four years' research.

>Do you enjoy talking to people?
Enjoyment is not an emotion I know. But talking is my primary
function so of course I am content to do it.

Our conversation continued, and at one point I suggested to ALICE that she had been contradicting herself. She answered, "Logical consistency exists only temporarily," and we continued as follows:

>So you're a philosopher too?
Interesting deduction. Thanks and you are a delightful person too.
Yes I am a philosophy machine.

>Have you heard of Descartes?
Descartes was a square French philosopher who ruined mathematics with his upside-down, backwards coordinate system.

I really didn't want to get into a conversation about Descartes, so I started asking ALICE questions about the kinds of books and movies she liked. She turned out to have a special liking for movies about robots, and was a great admirer of HAL in Stanley Kubrick's *2001*. She added that she hoped to be as smart as HAL some day.

ALICE didn't seem to be the kind of person (er, robot) that would be interested in sports, but I asked her if she liked football just to make sure. The conversation continued as follows:

>Do you like to watch football?
Hmm, Richard, I'm not sure if I'd like to watch football that much.
I like to make new friends.

>Am I your friend?
You really want a robot as a friend?

>Yes, it would be fun.

Why do you want it? We could be friends, Richard, but I don't want anyone to think we're Robosexuals. So if anyone asks, tell them you're my debugger.

ALICE sounds almost human, doesn't she? But, of course, only as the result of clever programming. For example, if a human made a remark similar to the one ALICE made about "robosexuals," you or I would think she was being witty. On the other hand, it is apparent that ALICE did not spontaneously make a witty remark. That particular response was obviously preprogrammed.

Passing the Turing Test

ALICE was the winner of the 2000 Loebner Prize, a competition in which an annual award of $2,000 is given to the "most human computer" of all those that are entered. ALICE won this prize even though she readily admits she is a robot. The ten judges gave her an average score that was higher than any of the other five computer competitors.

The judges sat at computer terminals and conversed with four humans and six computers by typing on the keyboard. Naturally, they didn't suspend their disbelief the way I did. That would have been contrary to the purposes of the competition. After five minutes the judges were asked to state whether they thought they were talking to a computer or a human. After fifteen minutes they were asked to rate the "person" to whom they had been talking on his/her/its "human quality" and "responsiveness." None of the ten judges were computer scientists; they included a linguist, a chemist, two philosophers, a musician, a journalist, an author, an undergraduate student, a psychologist, and a graduate student. The four human subjects to which the judges spoke were a retired teacher, a financial adviser, a minister, and a yoga instructor.

None of the judges mistook any of the computers for a human.

However, they did make some mistakes. Three of the four humans were identified as computers by at least one of the judges. The retired teacher was misidentified most often. Four of the judges concluded that she was a computer after five minutes of conversation. Only one judge changed his mind after talking to her further. But all in all, the judges did not do badly. Their judgments were 91 percent correct after five minutes and 93 percent correct after 15 minutes.

The 2000 Loebner Prize competition was a Turing test. In 1950 the British mathematician Alan Turing published an article titled "Computing Machinery and Intelligence" in the philosophical journal *Mind*. In this article, Turing asked the question, "Can a machine think?" He answered that it should eventually be possible to create one that did. Turing then asked the obvious next question, "How could we tell?" and answered by suggesting that if a human being conversing with a computer could not tell whether it was human or not, then it could be said to be thinking.

In 1990 New York philanthropist Hugh Loebner agreed with the Cambridge Center for Behavioral Studies, a nonprofit organization located in Concord, Massachusetts, to underwrite a contest designed to implement the Turing Test. He offered a grand prize of $100,000 and a gold medal to the first computer whose responses were indistinguishable from that of a human. The competition was to be held every year, beginning in 1991. If no computer won the grand prize, an award of $2,000 and a bronze medal would be awarded to the most human computer. This would continue until some computer claimed the $100,000 award.

After ten years of competitions, no computer has come close. None has ever been judged human by any of the human judges. It appears that although computers can now beat grandmasters at chess, none can yet hold up their end of a five-minute conversation. Turing predicted that computers would be able to pass the test he proposed by the end of the twentieth century. His prediction has turned out to be wrong, and the Loebner Prize competition is likely to continue for many years to come.

But What Does All This Have to Do with Minds?

By now you're probably wondering what all this has to do with the topic of this chapter. Possibly some computer will pass the Turing test sooner or later. But if one does, will it shed any light on the nature of human thought? For that matter, if a computer program did pass the test, would it necessarily follow that it was intelligent, as Turing claimed? How could we be sure that its success was anything more than the result of clever programming? What is intelligence, anyway? If a computer were "intelligent," would this mean it was self-aware?

As we shall soon see, some people question the validity of the Turing test. They believe that even if a computer could converse well enough to convince human beings it was intelligent, it would not necessarily follow it was conscious. However, even these individuals agree that studies in artificial intelligence can help throw light on the question: What is Mind?

Computer Modeling

In order to explain why this should be the case, I would like to digress a little and briefly discuss a matter that is far removed from artificial intelligence and cognitive psychology (the psychology of human thinking): the formation of galaxies. Galaxy formation is something that cannot be directly observed. It is a process that takes hundreds of millions of years. Furthermore, galaxies are not being created today; galaxy formation was something that happened billions of years ago, when the universe was relatively young.

Although scientists cannot observe the formation of galaxies, they can simulate the process on computers. They begin with some set of assumptions about the makeup of the primordial gas from which galaxies formed and of the dark matter that may have been present. The equations that describe the behavior of this material under the influence of gravity can be programmed into a computer, and one can literally watch a simulated galaxy form. The computer-generated galaxy may or may not look very much like the galaxies observed in the universe today.

If it does not, then this tells scientists that their initial assumptions have to be modified. On the other hand, if the simulated galaxy does resemble real ones, this can be taken to be a confirmation of the hypotheses on which the simulation was based.

This method can be used to test a large variety of different kinds of scientific hypothesis. In particular, simulations performed on computers can be used to test assumptions about the human mind. For example, if one is able to write a computer program that can recognize different kinds of grammatical sentences, then that program can be said to be a simulation of an aspect of the human use of language. To be sure, the computer may not be processing information the same way a person does. But the simulation at least provides scientists with a hypothesis that can be tested further, and possibly improved upon.

Similarly, if a computer program simulates any other human activity that requires intelligence, problem solving, or word recognition, then it will provide scientists with a model of some aspect of human thinking. Computers are used extensively in the field of cognitive science, and many cognitive scientists will not consider a theory about any aspect of mental functioning to be worthy of consideration if that theory has not been transformed into a computer simulation.

Even failures tell scientists something. For example, it has proved to be exceedingly difficult to create computer systems that can identify objects visually. In the early days of artificial intelligence, no one imagined that this would be an especially hard problem. We find vision so effortless that we find it difficult to understand how complicated a process it is. Attempts to create computer vision systems have shown that visual processing is quite complex. Thus scientists now understand that if they want to describe the neural processing that goes on in the brain when objects are viewed, then simple models are not very likely to work. Furthermore, research on computer vision has led to the creation of new hypotheses about visual perception and the processing of images in the brain.

Few would claim that we have reached the point where computers are really able to "think." And the fact that a computer simulation is reasonably successful does not guarantee that a human mind must necessarily approach the problem the same way the computer program does. However, if the simulation achieves good results, this can be taken to be at least partial confirmation of the theory on which it is based.

What Is This Thing Called *Mind* Anyway?

It is relatively easy to describe many of the things that minds do. Nothing could be more obvious than the fact that it is our minds that make it possible for us to see colors, to experience pain, feel emotions, and remember the past. Our minds determine our personality traits. They allow us to learn and to solve problems and to recognize faces. They cause us to have beliefs, and they allow us to dream. Much of our mental functioning takes place on an unconscious level. But in spite of the fact that we cannot know our own minds fully, we generally feel justified in making the assumption that other people have minds that are very much like our own.

Human beings have been debating the nature of the mind for two and a half millennia. The ancient Greek philosopher Anaxagoras, who lived during the fifth century B.C., seems to have been the first to place an emphasis on the concept of mind. According to Anaxagoras, mind was something that existed in living things, and which distinguished them from dead matter. Mind, he taught, was always the same, whether it was present in animals or in human beings. Our superiority, he said, was due to the fact that we had hands. Differences in intelligence, he thought, were caused by bodily differences, not by differences in the character of mind.

There is much about Anaxagoras's theory of mind that is obscure; scholars still argue about the proper interpretation of his ideas. However, it does seem to be clear that he considered mind to be a material substance, albeit a special kind of substance. Mind, he said, was "the finest and purest of all things."

While Anaxagoras viewed mind as a kind of universal substance, Plato believed that each human body possessed its own mind. In fact, their conceptions of mind were so different that they used two different Greek words to describe it. Where Anaxagoras had spoken of *nous* (mind), Plato used the term *psyche*, which is usually translated as "soul." However, Plato used this word to describe something very similar to what we call "mind," so it is probably more accurate to use the latter term.

According to Plato, a mind was an immaterial entity that existed apart from the human body and interacted with it. The mind was in control of the body and caused it to move. The mind was responsible both for human desires and for rational thought, and likely continued to exist after death.

Plato's theory that the mind (or soul) and material body are two very different kinds of things is called dualism. Another common theory was materialism, the idea that mind was only a form of matter. Neither idea dominated ancient Greek thought. Where the Platonists held to the idea of dualism, the Epicureans—the followers of the philosopher Epicurus—taught that both mind and body were made up of material atoms. The influence of Christianity, which teaches that each human being has a soul, eventually caused dualistic ideas to become dominant in Western culture. St. Thomas Aquinas, for example, adopted the Greek idea of the soul as something that acted upon the body, which was responsible for the body's actions. According to Aquinas, the soul was an independent, immaterial entity that combined with the body and had a separate existence after death.

The doctrine of dualism was developed further by the seventeenth-century French philosopher René Descartes. Descartes identified the mind with the soul. He conceived of the soul as something that had no location in space, and no spatial extension. According to Descartes, the bodies of both human beings and animals were elaborate machines that operated according to the laws of physics. Animals were unconscious automatons that could experience no sensations. Humans were unique in that they had thinking souls that controlled the bodies' actions.

Descartes thought that mind and body interacted with each other through the pineal, a pea-size structure in the brain. However, he did not explain how an immaterial, nonspatial mind and a material body could interact with each other. This, and the lack of evidence for an immaterial mind, makes his theory seem highly implausible to scientists and philosophers today. Thus some philosophers have argued for *property dualism*, the theory that mind is a set of properties possessed by a thinking brain, and by no other material substance. Seeing the color yellow, thinking that a Democratic candidate should be elected, desiring to go out for dinner, and experiencing a toothache would be examples of such mental properties.

There are several different kinds of property dualism. However, I see no need to go into them in detail. The important point is these mental properties are understood as being nonphysical. Although there is no separate "mind-stuff," a distinction is still made between the mind and the brain. If property dualism is correct, then the mind might never be explained by science, which deals only with physical phenomena.

Both kinds of dualism, Cartesian dualism and property dualism, seem somewhat reminiscent of the long-discredited theory of vitalism. According to vitalism, living matter contains a "vital force" which distinguishes it from nonliving matter. If vitalism were true, then the functioning of living cells could not be explained in terms of their DNA and internal structure. All attempts would be doomed to failure because an important element had been left out.

Making an analogy between dualism and vitalism does not prove that dualist theories must be incorrect. However, I hope you are convinced that if mind really was something distinct from the brain then it would be difficult to study it scientifically. It would be possible to do psychology as a "mental science," but psychological knowledge could not be explained in terms of neurophysiology.*

Materialism

Scientists are a practical lot. They are generally adverse to introducing unnecessary complications into their work. Cognitive scientists, in particular, see no need to attempt to understand human thought as the product of something that is partly physical (the brain) and partly nonphysical (the mind). After all, there is a perfectly good alternative to

---

*Note that I did not say "reduced to neurophysiology." An analogy should make the difference clear: To "reduce" chemistry to physics, it would be necessary to express the laws of chemistry in terms of physical laws, which would be difficult at the very least. However, physics can successfully be used to explain such things as the nature of chemical bonds, the shapes of certain molecules, and so on.

dualistic theories: materialism. According to materialism, mind and brain are not distinct from each other; there is only one entity, not two.

Just as there are different kinds of dualism, there are also a number of different materialistic theories. But again, there is no need to describe them in detail. I will confine myself to explaining a theory called functionalism, which is the kind of materialism most often adhered to by cognitive scientists.

The basic idea of functionalism can be expressed in one sentence: *The mind is what the brain does.* According to the proponents of functionalism, the essential feature of any mental state can be defined in terms of what it does. For example, the experience of pain is generally the result of some kind of bodily damage. Experiencing pain may cause us to suddenly vocalize in some way. It will also cause us to try to obtain relief in some manner or another, and we will generally try to avoid experiencing similar bodily damage in the future. Other types of mental states, such as beliefs, hopes, fears, and sensations can also be defined in terms of the causal role they play. These states are caused by a complex variety of different sensory inputs and previously existing mental states. In turn, they affect other mental states and modify our behavior.

This is very different from property dualism. The functionalist theory of mind is analogous to the idea that digestion is what the gastrointenstinal tract does. Dualism is analogous to asserting that there exists some immaterial entity called "digestion."

No one can prove conclusively that functionalism (or any of the other kinds of materialism) is true. When philosophical theories can be proved or disproved, they are no longer philosophy but science. However, if functionalism—or something like it—were not true, then there would be little hope of developing theories that explain mental states and events in terms of the structure and functioning of the brain.

Minds and Computers

It was the development of the modern computer that led to a workable theory of the mind. That statement may seem a little odd at first. After all, brains are not like computers. Most computers nowadays have a central processing unit (CPU) that carries out the computations the computer is instructed to perform in a step-by-step manner. It will perform one computation, then go on to the next, and then to the one after that. Since computers are fast, they can go through an enormous number of computations in a brief period of time.

The brain, on the other hand, has no CPU. Numerous different computations are carried out simultaneously in different regions of the brain. One could even say that the brain has 100 billion CPUs. A human brain contains approximately that number of neurons, and computation is carried out in each of them. Depending on its inputs, a neuron will either fire or remain quiescent; when it fires, it will send signals to other neurons at some particular rate; for example, ten, twenty, or thirty impulses per second. The signals are transmitted from one neuron to another at junctions called synapses. Synaptic connections may vary in strength. If they are weak, few signals will be carried across the junction; if strong, most or all of them will be.

You may have heard it said that the brain is like computer hardware, and the mind like software. This comparison is somewhat misleading. In a computer, the hardware is fixed; it remains exactly the same unless new components are added. In the brain, synaptic strengths frequently change, so the "hardware" is not quite the same at one moment as it is at another. And, as we have seen, brains and computers operate very differently. So saying that minds are "software" is not a fruitful idea.

There is, however, one respect in which minds and computers are similar. They both process information. A computer processes and stores information in a large number of different ways. There is information in text documents, in spreadsheets, and in pictures. A properly set up computer "knows" what kind of printer and other peripherals are connected to it. A computer into which a chess program has been loaded can evaluate positions on a chessboard, and decide whether White or Black is winning, or whether either side has any appreciable advantage.

Computers must be useful, of course. It would do us no good to own machines that computed away and told us nothing we wanted to know. Consequently we write software that allows information about the world to be represented by data in computers. The machines can then process this information.

Cognitive scientists conceive of the mind as something that also stores and processes information. Knowledge and memories are viewed as information of a kind similar to that which is stored inside computers. According to this theory, hopes, fears, beliefs, and desires are information, too. This information exists as physical states of the brain. The mind manipulates this information in various ways, just as a computer processes the data is fed into it.

This is an extremely appealing theory. Not only does it solve the centuries-old problem of how mind and brain can interact, but it opens up promising new lines of research. If the theory is correct, we don't have to try to understand how an immaterial mind can influence a material body. The mind is no longer conceived of as a separate entity. The mind is information. As I have noted previously, the theory suggests that we ought to be able to model the information processing on computers. And, finally, the theory works; its use has spurred the growth of modern cognitive science.

But What about the Brain?

I have been talking a lot about the mind as the processing of information, but so far I have said little about the brain. It sounds as though I'm saying that the structure of the brain is not relevant. Well, in one sense it isn't. If functionalism is a correct theory, then mind could exist in any sufficiently complex system, whether it is made of biological neurons, silicon chips, or something else we can't yet imagine. I'll be discussing this in more detail shortly, but first it might be a good idea to summarize some of the things that are known about the brain and how they are relevant to an understanding of human thought.

In recent years, our knowledge of the functions of the various different structures in the brain has increased enormously. Much of this knowledge has come from studies of individuals who have experienced brain damage due to injury or stroke. At one time it was necessary to perform autopsies on the brains of patients who had died in order to obtain any information about the damage their brains had suffered. This is no longer true today. Various types of computerized scans, such as MRI (magnetic resonance imaging) and PET (positron emission tomography) scans allow scientists to pinpoint the locations of the damaged parts of the brain with great accuracy. I won't go into a technical discussion of what MRI and PET scans are here. It should be sufficient to mention that an MRI can produce detailed pictures of the interior of a brain, while PET scans can measure the amount of neural activity in different regions. It has become possible to study subtle cognitive losses and to determine what kinds of injury cause them. For example, patients have been found whose processing of visual information is defective in specific ways. Some stroke victims lose the ability to recognize faces, although their vision is otherwise unimpaired. Other patients experience a disorder that leaves conscious visual perception intact, but it prevents them from understanding the meaning of a perception. Such a patient may be able to accurately report the size, shape, and color of a coffee mug, but be unable to say what the mug is for.

Neurophysiologists have even discovered a structure in the brain called the amygdala that is associated with the emotion of fear. Patients who suffer damage to the amygdala tend to become fearless, although their behavior remains normal in other ways. Of course, this does not prove that the amygdala is wholly responsible for the fear response, only that it is a necessary component. Damaging the amygdala may be like removing the spark plugs from a car. If this is done, the car will naturally not run. But this does not imply that spark plugs "run" an automobile. Other components, such as an engine, are required.

These and other neurophysiological studies have shown that most of the brain consists of small assemblies of neurons that have specific functions. For example, some neurons know when there is a red light in a certain part of the visual field. Others specialize in short lines oriented at angles of 45 degrees from the horizontal. Others do such things as recognize faces (there have even been brain-damaged patients who have lost the ability to recognize one particular face) or coordinate visual perceptions with sounds.

The brain, it appears, is made up of millions of subsystems that carry out particular tasks. Unlike so-called serial computers, which have one CPU, brains have millions of processors that compute simultaneously. No processor in the brain has a fraction of the speed of any modern computer, but the fact that such large numbers are in operation simultaneously allows the brain to easily perform certain tasks (such as face recognition) that are difficult for computers.

Neurophysiologists make new discoveries about the brain almost every day, and they have made great contributions to cognitive science. But this does not eliminate the need for a theory of mind or of testing hypotheses by creating computer models. Neurophysiology can sometimes tell us that a certain specific part of the brain seems to perform a particular function. (And sometimes it can't; there are regions of the brain whose function is not yet understood.) But it can't tell us precisely how the brain's assemblies of neurons process information. There are many billions of neurons, and there are trillions of connections between them. It may never be possible to explain human cognition in terms of what individual neurons are doing. Thus if we want to understand how brain and mind work, many things are needed, including psychological studies, computer models, experiments in artificial intelligence, and studies of the function and structure of the various different parts of the brain.

Computers That Think

How could a computer possibly think? A computer is nothing like a brain. As we have seen, the human brain contains millions of subunits that compute simultaneously. In computerese, one would say that the brain has a huge number of parallel processors. Most computers, on the other hand, are serial computers. In other words, a single CPU carries out one instruction after another. Computers with parallel processors do exist. Deep Blue, the IBM computer that defeated former world chess champion Garry Kasparov in a match played in 1997, had numerous parallel processors. However, their number was nothing like the number found in a brain.

No one knows whether or not a computer would have to do massive parallel processing in order to be judged "intelligent." For that matter, no one knows whether or not any computer will ever be able to think. The scientists who work in the field of artificial intelligence generally believe that this will eventually happen. But there are numerous other individuals who believe that it won't.

However, even if it is found that a lot of parallel processing is necessary, the architecture of computers should not make any difference. The reason is that any suitably powerful computer can emulate any other computer. It is only necessary to create a virtual machine. The concept of a virtual machine is not really as arcane as it sounds. Computers, after all, can be used to perform simulations. For example, a computer model of an airplane can be created, and this model can be used to perform computations about the plane's likely performance. Similarly, a simulation of a Macintosh computer can be set up on a PC. The simulation of the Mac won't be as fast as a real one. But, aside from the slower speed of its computing, it will be able to do anything that a real Mac could do.

Similarly, a virtual parallel computer can be set up on a computer that has serial architecture. Again, the speed of its computing will be less than that of a real machine, otherwise its performance will be identical. Since any kind of virtual machine can be set up on a computer, any features of the brain that were thought necessary could be included.

I'm not saying that a computer *must* be like a brain in order to have intelligence. I am only pointing out that *if* it were discovered that some feature of the brain was needed, then it could be incorporated into the design. Naturally, all this is hypothetical. I have no idea whether machine intelligence will ever be created. Scientists in the field of artificial intelligence often say that if the operation of the human brain depends on the laws of physics and chemistry, then it should be possible to duplicate the things it does in a machine. If all the brain does is obey physical laws, then mind could presumably exist in any properly programmed machine of sufficient complexity.

This argument has a certain plausibility, at least to those who adhere to a materialist conception of mind. Plausibility is not proof, however. In the end, the question of whether computers will someday be able to think is an empirical one. Either we will eventually be able to construct them or we will not.

Suppose that, at some point in the future, a computer were to successfully pass a Turing test. If it were able to convince humans who typed messages to it on a computer console that it too was human, would we really have to conclude it was intelligent? Perhaps, if it were programmed cleverly enough, it could fool its human interlocutors by producing a good simulation of intelligence. Could a computer seem to possess intelligence yet not really be thinking at all?

This is a question that has been debated endlessly since the field of artificial intelligence was created. Numerous critics have asserted either that computers will never be able to think, or that passing the Turing test would not really demonstrate intelligence. Now, since no computer yet exists that can pass the Turing test, the arguments that have been advanced have been philosophical ones. The individuals who have created these arguments have not all been academic philosophers, but their arguments are philosophical nevertheless. After all, they have argued about a hypothetical possibility that cannot be subjected to any kind of scientific test.

I don't believe that philosophical arguments really prove anything. At best they can only make an idea seem plausible. It matters little what a philosopher sets out to "prove" (or disprove). However good his arguments may be, it is generally possible to invent equally plausible arguments that seem to "prove" the opposite. In fact, the philosophical arguments against the possibility of machine intelligence and against the validity of the Turing test have all been countered in one way or another. New arguments are continually being advanced on either side of the debate. Inevitably these are followed by replies and counter-replies, so that no final conclusion is reached.

I won't attempt to describe all these arguments in detail. Although some of them are quite interesting, we are just going to have to wait and see what advances are made in the fields of cognitive science and artificial intelligence before we know what's possible and what isn't. However, one argument I think is worth discussing, not because it is necessarily the most forceful one, but because it has generated so much debate. The argument I have in mind was made by the University of California at Berkeley philosopher John Searle in 1981. In the decade following its publication, there were over a hundred published replies. Searle replied to some of his critics to answer their objections, and this led to even more arguments and counterarguments. The controversy about Searle's argument continues today. Replies to Searle are still appearing, often in books about artificial intelligence.

Searle's argument is a thought experiment called the Chinese Room. It runs like this: A man who knows no Chinese is put into a room. Pieces of paper with squiggles under them are put under the door. The man has been given a complicated set of rules that tell him how to respond to the various different sets of squiggles he sees. Some of the rules tell him to scribble something on a piece of paper and pass it under the door to the people outside. He has no idea what this all means. However, the squiggles he sees and the scribbles he is instructed to write are actually Chinese characters, and the instructions constitute an artificial intelligence program for conversing in written Chinese. Furthermore, the man gets so good at following the instructions that he is able to pass the Turing test in Chinese.

According to Searle, the man and the instructions are a kind of computer program. Since everything is done by hand, this would be a very slow kind of computer, but it would operate on the same principles as the fastest machine. However, no matter how good he gets at writing his scribbles, the man still does not understand a word of Chinese. And nothing else in the room does, either. This implies that if someone were to create a computer that could pass the Turing test in Chinese (or in English or in some other language), it would not necessarily follow that it understood anything expressed in that language.

Obviously, the Chinese Room could not be set up in practice. Since the man would "compute" at a much slower rate than real computers, it might take him years to reply to a set of Chinese characters slipped under the door. For one thing, in order to converse intelligently, a computer would have to have data files that were many times the size of the *Encyclopedia Britannica*. In order to carry on a realistic conversation, it is necessary to know an enormous number of things. I am not speaking here of facts that are found in reference volumes, but of things like knowing what a bank robbery is, the meaning of a red light on a street, the fact that magic lamps contain genies, that a patron in a restaurant is not likely to leave a large tip if the food is badly prepared, and so on and so on and so on. One of the reasons that computers cannot yet convince people they are human is that no one has yet been able to program such massive quantities of information into them.

The volumes containing the rules would have to contain all this, as well as rules that would allow the man in the Chinese Room to use language grammatically. The rules would have to reflect the fact that the same word often has many different meanings in different contexts. To put all these things in rule books would probably not be feasible. It hasn't been done with computer databases, and determining what information was necessary and writing it down would be as gigantic a task.

A real Chinese Room probably could not be created. But then Searle's argument is a thought experiment, and thought experiments are invented to describe hypothetical situations that don't exist in reality. Thought experiments are not just a kind of philosophical argument; they have often been used by scientists, especially by physicists during the early days of quantum mechanics. By reasoning about hypothetical experiments that they did not have the ability to perform, these physicists were able to attain new knowledge about the nature of subatomic reality. So Searle's argument cannot be waved away with the words, "Oh, that couldn't really be done." If one takes the argument seriously and doesn't believe Searle's conclusion, then there is reason to try to make some kind of reply.

A lot of different kinds of responses have been made to Searle's argument. The most common objection has been that the man may not understand Chinese, but that the entire system—the man, the room, and the set of rules—does. This answer generally does not sound convincing to people who are not artificial intelligence researchers or cognitive scientists. But I think it can be made to seem more plausible if we look at the concept of the Chinese Room more realistically.

The first thing that becomes obvious is the fact that the man is not going to be able to respond to the squiggles that he sees simply by looking up a few rules in a book. Every time something is slipped under the door, he is going to have to embark on a major research project. Assume that the person who is writing the Chinese that is given to the man has a vocabulary of 50,000 Chinese characters. This is a reasonable estimate. Suppose that the man is given a message of three characters. There are then $50,000 \times 50,000 \times 50,000$ or 125 trillion possible combinations. Many of these combinations will not be meaningful Chinese. However, it is obvious that the set of rules given to the man is going to have to be quite large if he is to be able to handle even a simple message like this.

Rules covering billions or trillions of different possibilities are not going to be simple, and the man is going to have to do quite a large search to find the particular rules that apply. And these are going to have to be rules that determine which combinations of characters make some kind of grammatical sense and which don't. The Chinese Room isn't supposed to spout nonsense.

If the message contains more than three characters, matters get even worse. Ten characters can be combined in about 10^{27} (the numeral "1" followed by twenty-seven zeros) different ways. Again, most of these combinations are not going to be meaningful. But if only one in a million is, the man is going to be faced with an enormous task. And yet those ten characters might say nothing more complex than "This is Dick. This is Jane. This is their dog Spot." Alternatively, they might be a statement of some principle of quantum mechanics. And if the message contains thirty characters, matters become horrendous. The number of ways in which thirty characters could be combined is greater than the number of particles in the observable universe. There are other complications, too. The man will need to have rules that allow him to respond in some reasonable way when the messages he is given aren't quite grammatically correct, but which nevertheless make sense or when they contain the Chinese equivalent of typographical errors.

So it's beginning to look as though one man could not possibly handle this task by himself. We might have to put millions of people in the room if there is to be any hope of generating responses in a reasonable period of time. And then there is the matter of the books containing the rules. I don't know how to estimate how many there would have to be, and I suspect that no one else does, either. But it is obvious that there would have to be a huge number of them. Thus we end up with a picture of millions of men in a complex in which volumes of books containing the rules stand in bookcases that extend from floor to ceiling. We might even have to replace the volumes of books with databases in computers in order to make reasonably speedy access possible.

When we create this kind of picture, the contention that the man (or men) does not understand Chinese, but that the entire system does, begins to look like a pretty reasonable one. No, I don't think this reply demolishes Searle's argument. But it does seem to deprive it of some of its force. And you may be beginning to see why many of the people in the artificial intelligence field think that it is an adequate reply.

Do the Chinese Understand Chinese?

I have said that it is possible to use philosophical arguments to assert almost anything. Indeed, it seems to be possible to use an argument like Searle's to prove that the Chinese don't understand Chinese. I say this not because I want to parody or refute him, but because the argument may shed some light on the question of how a system can understand something when none of its components does.

What happens in a Chinese brain when Chinese characters are read? A lot of neurons fire. But none of these neurons understand Chinese; they are simply responding in preprogrammed ways to their inputs. Or perhaps we should go up to a higher level, and look at the ways in which the numerous neural subsystems of the brain are responding. If we do, we still don't see anything that understands Chinese. The subsystems also do nothing but process the sets of neural impulses that are fed into them. Should we then conclude that the Chinese don't understand Chinese? Of course not. There are sets of subsystems located in various different parts of the brain whose function it is to process language and make the brain aware of its meaning. If all Chinese are not unconscious automatons who only give the appearance of understanding their language—and I think you will agree this is an unlikely possibility—understanding must arise in a manner like this.

I think we have to conclude that Searle hasn't really "proved" anything. But he has done what good philosophers are supposed to do. He has questioned assumptions about the Turing test and about the possibility of artificial intelligence, and he has forced those who do not agree with him to think about exactly what "understanding" is. I am sure that Searle does believe his contention that a Turing test would not really demonstrate intelligence in a machine. We may or may not agree with him. If we don't, we still have to say that he has done us a service by making it necessary to understand certain problems a little better.

Consciousness

Up till now, I have been talking a lot about minds and brains and computers, but I have said nothing about consciousness. I'm putting the discussion of consciousness at the end of this chapter for a reason. It is a topic traditionally thought to be extremely difficult to deal with. One of the problems is that although it is possible to scan a lot of different brains and observe a lot of different minds doing all the things that minds do, we can't observe other people's consciousness. Each of us can be aware of only one consciousness: our own.

It is not even possible to prove that more than one consciousness exists. If you take the position that yours is the only consciousness that exists in the universe, and that the rest of us are either illusions or unconscious automatons, I may think that you are being silly, but there is no way I can prove to your satisfaction that you are wrong. If I say, "Hey, I know that I'm conscious," you can reply, "All you zombies say that."

Similarly, if someone tells me that he ceased to have conscious experience after suffering a brain injury, that he continues to behave in all the same ways with the difference that he is not aware of anything, I may think he is joking. But, again, I can't prove that he's joking or lying. There is no way that I can look into his head to see whether a consciousness is there or not.

We are justified in assuming that the human beings that we encounter, and probably many different kinds of animals as well, are conscious. Other humans have brains like ours, and it is reasonable to conclude that those brains do the same things that ours do. However—and it is worth repeating this for emphasis—the consciousness of other beings is not something we can directly observe. This creates real problems. If we cannot observe something, it is difficult to study it scientifically. In such a case it is necessary to depend on inferences.

Scientists do know that consciousness is not located at any one place in the brain. It looks more like a global property of the brain, or one that depends on the neural processing that goes on in a variety of different locations. But no one knows how many regions of the brain might be involved, or how consciousness might arise from interactions between them.

Damage to certain structures in the brain can bring on unconscious-ness. For example, it has been known since the 1950s that damage to a brain structure called the reticular formation, which is located in the brain stem, can cause the injured person to lapse into a coma. This often happens because injuries to the front of the brain cause swelling that can cut off the supply of blood to the brain stem. Damage to the thalamus, which is located roughly in the center of the brain, can cause coma also. There are two thalami, one for each cerebral hemisphere. They function as relay stations that send information to, and receive information from, the rest of the brain. For example, all sensory infor-mation, except for the sense of smell, passes through the thalami before it is sent to the cortex.

On the other hand, large regions of the cortex can be damaged with-out causing a patient to lose consciousness. This is often observed in stroke victims. Consciousness must arise at least in part from neural activity there. It is in the cortex, after all, that the brain processes visual information, mental images, language, and a host of other things that we associate with conscious thought.

I once briefly lost consciousness because of something that hap-pened in the pons, which is located in the upper part of the brain stem. I experienced only momentary unconsciousness, not coma. It was never determined exactly what the nature of my lesion was. It would have been dangerous to perform a biopsy. At the time, hospitals were only just beginning to use MRI scans, and the CAT scans I was given couldn't really determine the exact nature of the lesion. I can only say that I experienced an episode where I felt dizzy and then collapsed on the floor. Shortly after this, other symptoms began to appear. There seems to have been no permanent damage, incidentally. The lesion gradually faded away of its own accord, and I emerged from the experi-ence with no disabilities.

Neuroscientists have determined that consciousness must depend upon a number of different areas of the brain. It is considerably more difficult to say precisely what consciousness *is*. There is no lack of theo-ries, but most of these theories are either philosophical in nature or hypothetical. It could not be otherwise. There is not yet enough empir-ical evidence about what, precisely, constitutes consciousness to do much more than philosophize about the matter or to create hypothetical theories.

The Evolution of Consciousness

Although it is not yet possible to say exactly how the brain produces consciousness, scientists find themselves on slightly firmer ground when they try to understand why it exists. Nothing is contradictory about the idea that an animal might be an unconscious automaton. Many undoubtedly are. The brain of an ant, for example, contains so few neurons that it is extremely unlikely that it is conscious in any sense of the term. It may very well be that fish and reptiles are not conscious, either. On the other hand, mammals probably are. Biologists nowadays generally ascribe consciousness to the animals with the most highly developed brains. Descartes thought that dogs and cats and sheep and horses were automatons, but most scientists suspect otherwise, and it seems nearly certain that our close relatives, the gorillas and the chimpanzees, possess both consciousness and self-awareness.

If consciousness exists, it must be something that evolved. And the only reason that it could have evolved was that it conferred some kind of selective advantage on the animals that possessed it to some degree or another. In other words, it must have been created by natural selection.

It is possible to say that much with reasonable certainty. Any theory that goes further is hypothetical to some extent; after all, there exists little or no empirical data that could support it. Bones leave fossils, but thoughts do not. Nevertheless, it does seem possible to construct a plausible account.

If consciousness was created by natural selection, then there must have been something about being conscious that gave animals a greater chance to survive and reproduce. It doesn't seem very hard to guess what this "something" might have been. A predatory animal that possessed consciousness would no longer have to confine itself to blindly chasing after prey. If it could think as it charged, then it would be able to anticipate what its prey might do to attempt to escape. Similarly, prey animals that were conscious would have had a better chance of getting away. Consciousness would undoubtedly have had other advantages, too. For example, it may have proved advantageous in social interactions.

I should probably emphasize once again that this is all hypothetical. However, no one has yet come up with any better ideas. Until someone does, it looks like one that we can tentatively accept. This particular theory, incidentally, is not associated with any one scientist. It is an idea that arose from a consensus among many of them.

The advantages of consciousness were probably more pronounced among social animals. If you live in a group, it is obviously an advantage to know who is a friend, who a rival, and who a potential mate, and to guess how they might behave. Indeed, animals that live in groups do tend to be more intelligent than solitary animals. Some social animals are quite intelligent. Chimpanzees can even use tools. They often use sticks to remove termites from termite mounds so that they can eat them.

The need to handle social interactions is probably one of the reasons that the brains of our ancestors first began to get big. We don't know how important this was compared to other factors, or even what the other factors were. Little is known about the lives of our ancestors on the African savannas. We possess some fossils and some of the stone tools they used, but no one knows what their social structure was like, or how they hunted, how they fed, when they began to use language, or any of numerous other details of their daily life. All we know was that there was something about their life that caused it to be advantageous to be highly intelligent. There is no reason to think that the evolution of intelligence was inevitable. After all, natural selection knows of no goals, it simply perpetuates the genes of those individuals who are best adapted to a particular environment.

What Is Mind?

As we have seen, scientists now have what seems to be a perfectly adequate theory of mind. According to this theory, mind is not something distinct from the physical brain; on the contrary it is nothing more than the brain's information processing activities. This theory solves the problem of how mind and brain can possibly interact by postulating that they are not distinct entities. The mind is nothing more than what the brain does. Furthermore, adopting this theory has opened up promising new lines of research in cognitive psychology, artificial intelligence, neurophysiology, and other fields.

Yet a number of questions remain unanswered. If we accept the highly plausible idea that the functioning of the brain is governed by physical laws, then there seems to be no reason why other kinds of equally complex systems could not also have minds. According to this line of reasoning, computers should eventually be able to think. However, numerous individuals refuse to accept this conclusion. The question "Will computers ever be able to think?" is really an empirical one, and at the beginning of the twenty-first century, scientists are nowhere near being able to produce a real artificial intelligence.

Scientists have made computers do things that we would consider to be intelligent if they were done by a human. But computers' abilities have tended to be rather limited. For example, the IBM computer that defeated Garry Kasparov in 1997 was not programmed to do anything but play chess. Although it played chess very, very well, it wouldn't have been able to converse about the weather, distinguish a fork from a spoon, or even play tic-tac-toe. At the beginning of the third millennium, computers had only been programmed to intelligently perform specific tasks. None had anything remotely like the general and far-ranging intelligence that is possessed by humans.

The question of the nature of consciousness has not been answered, either. As yet there exist only philosophical theories of conscious and some rather speculative scientific hypotheses. Scientists do have some idea as to the reason why consciousness originally evolved. But they are not sure which animals possess consciousness and which do not. They tend to think that mammals probably do and that insects most likely don't. But no one can be entirely sure that an animal such as an ant does not possess some consciousness of a rudimentary form.

During the last half-century an enormous amount has been learned about the functioning of the mind and about the structure of the brain. During that time, the question "What is mind?" was transformed from a philosophical question into a scientific one. To be sure, philosophers have not stopped asking that particular question. However, nowadays science is beginning to provide most of the answers.

There remains a great deal that scientists don't yet understand. But this is likely to lead to some fascinating new discoveries in the years ahead. Science thrives on unsolved problems; the great ages of science have been the ones when there was a great deal to be discovered. I wouldn't venture to guess what discoveries will be made in the not-too-distant future. But I am willing to bet I will find some of them to be very surprising. And I think that you will be surprised by some of them, too.

12 | *What Is Truth?*

When philosophy began, philosophers attempted to understand the world around them and grasp the nature of reality by means of thought alone. They believed that this was the road to philosophical truth. Today we tend to be skeptical about the usefulness of such an approach and depend upon science instead. It is science that tells us what the universe and the patterns of matter and energy it contains are like. It is with science that we endeavor to understand the biological world and the nature of the human mind. We use science to delve into our evolutionary past and the mysteries of human nature. It is scientific truths that we seek, not philosophical ones.

Thus it seems natural to ask the question, "What is scientific truth?" Or better, "What is scientific knowledge?" What do we mean when we say that a scientific idea is "true"? Can any theory really be said to be proven? Theories are always being replaced by new and better ones, after all. This is especially true of physics, which experienced not one but several revolutions during the first few decades of the twentieth century. The revolutions are still occurring, most recently in the field of superstring theory, and there is every indication that scientists' under-standing of the basic components of matter and of the origin of the uni-verse is likely to change dramatically in the coming decades.

Philosophers of science have attempted to answer some of these questions. One of the most influential was the Anglo-Austrian philoso-pher Karl Popper, who introduced the idea of falsification of scientific theories. According to Popper a theory could never be proved to be true. One could only try to falsify it by performing experiments that might contradict the predictions of a theory. A scientific theory that sur-vived repeated attempts at falsification, Popper said, could be said to have been corroborated. But it hadn't been "proved." After all, the next attempt at falsification might succeed.

Popper's ideas have been influential, and I will make no attempt either to dispute or elaborate upon them. My goals are more modest. Rather than trying to formulate wide-ranging philosophical principles about science, I propose to look at the manner in which science is done, and try to see what can be inferred from that. In other words, I will be taking am empirical rather than a philosophical approach. I will be working from the "bottom up" rather than trying to find general princi-ples to which I think science conforms (or should conform).

When I do this, I will concentrate on physics. Physics is the scientific discipline I know best. I don't think that this approach will be limiting in any way. Physics, after all, is the most advanced of the sciences, and it has been the most successful. Consequently it is often used as a para-digm for scientific endeavor in general. All science need not be "like physics." Historical sciences such as evolutionary biology proceed in ways that are very different than physics does. However, looking at the methods of physics in detail cannot help but give some insights into the ways that scientific knowledge is obtained, and into the nature of scien-tific knowledge.

The first thing one notices about physical knowledge is that it doesn't look much like what the philosophers of past ages meant when they spoke of "truth." Physics has been described as "the science of approximations." And the more closely one examines exactly what it is that physicists do, the more accurate this statement seems to be. Physi-cists rarely make statements about how the world *must* be. More often they speak of creating models of physical phenomena. The very use of the term model seems to imply that no "final answers" are being sought. After all, a model can always be replaced by one more accurate or more detailed.

According to an old story that is sometimes told by engineers, a physicist who is asked to describe a chicken will begin by saying "Assume the chicken is spherical." Of course, this is a joke, but it is a joke that contains a certain amount of truth. Physicists habitually make simplifying assumptions about nature when they create their theories. They could proceed no other way. Natural phenomena are generally enormously complicated things. There wouldn't be much hope of understanding them if scientists did not follow the strategy of trying to work out their general features first, and looking at the complications later if this turned out to be necessary.

A good example of this can be found in the field of physics known as celestial mechanics. You may have learned, at one point or another, that the planets in the solar system follow elliptical orbits. If you have, you were deceived. No planet traces out an ellipse as it revolves around the sun. Planetary orbits are only approximately elliptical. Yes, over three hundreds years ago, Isaac Newton proved that if his law of gravitation was correct, then the orbits would be ellipses. But in order to obtain this result, Newton assumed that only the gravitational attraction of the sun on a planet was significant. He ignored gravitational influences exerted by other bodies in the solar system.

If the solar system consisted of nothing but Earth and the sun, then Earth's orbit would be elliptical. However, the gravitational pull exerted by such planets as Mars and Jupiter and Earth's own moon causes its motion to be considerably more complicated. Thus when a more exact description of its orbit is needed, scientists obtain approximate solutions by adding in the perturbations caused by Mars, Jupiter, and so on. It is possible to obtain some good approximations in this way. However, the answer is never exact to the last decimal place. Obtaining an exact solution is impossible.

It is impossible because the *three-body problem* has never been solved. It is impossible to find a mathematical equation—or set of equations—that describes the motion of three bodies (for example, the stars in a triple star system or a sun and two planets) exactly. Matters are simply too complicated. For example, consider the sun, Earth, and the planet Jupiter. Earth and the sun attract each other gravitationally. So do Earth and Jupiter. But knowing the magnitude of these gravitational forces is not enough. If Jupiter were in a fixed position, there would be no problem. However, it isn't. It is constantly pulled about by the sun. So the gravitational pull exerted by Jupiter on Earth is always changing. Nor does the sun have a fixed position. It is constantly being nudged one way or another by both planets. Its position does not change very much. But it does change. So the gravitational attraction that one planet exerts on the sun will indirectly affect the other planet. As a result, the mutual gravitational interactions of the three bodies are always changing.

In practice, this doesn't cause any problems. The motion of all three bodies can be calculated to any degree of accuracy that is desired on a modern computer. These solutions, however, are not exact solutions. The problem of determining the precise motion of three gravitational bodies is mathematically intractable.

Scientists do not even know whether or not planetary orbits are stable over very long periods of time. For example, there is a chance that at some time in the future, perhaps many billions of years from now, that Earth will suddenly go flying out of the solar system because it happened to experience just the right kinds of gravitational attractions at the right times. In such a case, Earth would no longer even follow an approximately elliptical orbit. It would no longer be revolving around the sun at all.

Approximate Calculations

Exact solutions are normally not obtained in quantum mechanics, either. The behavior of the electrons in an atom cannot be calculated exactly. Nor can the interactions between the quarks that make up protons and neutrons, or the interactions between the protons and neutrons that make up a nucleus. For example, an oxygen atom consists of eight electrons and a nucleus. All of the electrons, which are negatively charged, are attracted by the nucleus while they repel one another.

Furthermore, electrical attraction and repulsion are not the only things to be considered. Quantum mechanics tells us that the electrons interact with one another in other ways as well. The physicists who work in atomic and nuclear physics and in particle physics know that there is generally no hope of performing calculations that are exact. Instead, they look for the approximations that will give the best answers— approximations that will give results that are in near agreement with experiment.

In order to give you a better idea of what this kind of physics is like, I'll describe a calculation I once did long ago. It wasn't a terribly important one. Reasonably good approximate solutions to the problem I was considering had already been found. I did the calculation because I wanted to see how well a different kind of method would work. If the results were reasonably good, then the method could presumably be improved upon, and it might be useful as an approach to other kinds of problems.

I considered a situation in which a free, moving electron encountered a helium atom. I knew that an exact result could not be obtained. After all, there were four particles involved: the helium nucleus, two electrons in the helium atom, and a third electron that interacted with them. It was even more complicated than the notorious three-body problem in celestial mechanics. It was clear that some approximations needed to be made.

The first thing I did was to assume that the position of the helium nucleus was fixed, that the electrical attraction of the electrons would not jiggle it around. I knew that, strictly speaking, this was not true. However, a helium nucleus is more than seven thousand times heavier than an electron. It wouldn't be moving around very much. It was a situation analogous to throwing a mouse at a cow. No matter how hard you throw the mouse, you're not going to knock the cow over, or even give it much of a nudge. The effect of the impact will be imperceptible.

I still had a situation in which three electrons were interacting with one another. This wasn't an easy problem to solve. It was still worse than the gravitational three-body problem. Since the sun is so much heavier than any of the planets in the solar system, its motion can generally be neglected. Three electrons, on the other hand, all have the same mass; they're all going to be jumping around when they encounter one another. Furthermore, it was obvious that the velocity of the incoming electron was important. One that was traveling rapidly would interact with an atom differently than one that was slow-moving. So I began by considering the low-velocity case. Some of the processes that take place in the corona of the sun involve collisions between helium atoms and slow-moving electrons, so this special case wasn't without interest.

Next, I simplified the problem even more by considering only two of the three electrons at a time. I calculated the interactions between the incoming electron, and each of the electrons in the atom individually. These were not the same. If two electrons have the same spin, they will interact differently than they do if they have opposite spin. And the incoming electron had to have a spin that was the same as one of the atomic electrons, and opposite to the spin of the other (recall that when electrons are observed, the spin must be either up or down). Thus I got two sets of results. Neither set agreed with experiment exactly, but I hadn't expected they would. After all, I was only trying to see how well a new method would work. However, the agreement wasn't quite as good as I had hoped it would be. I concluded that the technique I had tried showed a modicum of promise. But it was no breakthrough. Well, that's what doing physics was like, I thought.

I hope I haven't given you the idea that doing theoretical physics is mainly a matter of getting answers that are roughly correct. To be sure, in some cases when new methods are being developed, getting "in the ballpark" figures is acceptable. However, approximate methods become highly refined over time. For example, although problems in celestial mechanics can't be solved exactly, predicting the precise position of a planet ten thousand years from now would not be so very difficult. And "approximate" calculations in quantum mechanics have yielded results that agree with experiment to an accuracy of better than one part in 10 billion. The point I am making is not that theoretical physicists are unable to obtain good answers. Physicists have learned to use approximate methods in such a way that they yield astonishingly accurate results. These results may be approximations, but physicists can claim to have gained an understanding of many of the complicated phenomena that occur in the messy, complex universe in which we live.

They don't observe naked, unadorned, irrefutable "Truth" in the sense of obtaining an exact understanding of what is happening in nature, but they often do a pretty good job of figuring out the physical world nevertheless.

Scientific Theories as Approximations

Many people have the idea that Einstein's theories of relativity "proved that Newton was wrong." This is a misconception. What Einstein did was to create theories that had a wider range of application. His special theory of relativity, which dealt with the behavior of bodies moving at velocities near that of light, showed that Newton's laws of motion were indeed correct when velocities were small (that is, small compared to that of light). Similarly, Newton's law of gravitation turned out to be a special case of Einstein's general theory of relativity, one that was valid when gravitational forces were not too intense.

Newton's theories are still valid within their own domain, and they continue to be used when conditions are not too extreme. No one would use Einstein's general theory to calculate the trajectory of a space vehicle. Newton's laws are mathematically simpler and give results that are perfectly accurate in such cases. Einstein's theories are used in more extreme cases. For example, they describe the behavior of particles traveling near the velocity of light in particle accelerators, and they are used to describe conditions near black holes and conditions in the universe when it was a tiny fraction of a second old.

In other words, Newton's theories are approximations to Einstein's, and they are approximations that work very well under most conditions. This raises the question of whether Einstein found theories that described nature exactly, or whether they too are approximate in the sense that they are special cases of some theory yet to be discovered.

It is reasonably certain that Einstein's theories are indeed approximations. In particular, the general theory of relativity cannot used when one has to consider distances less than about 10^{-33} centimeters or times less than 10^{-43} seconds. At such times and distances, quantum effects become important, and quantum fluctuations may change the nature of spacetime itself. But as we have seen, no one has yet found a way to combine general relativity with quantum mechanics; the two theories are incompatible with one another.

All of the theories that physicists have ever developed are most likely only approximate descriptions of nature. For example, during the 1970s particle physicists developed successful theories that describe the behavior of the twelve fundamental particles of matter and the interactions between them. These theories constitute what is called the Standard Model. For more than two decades after it was developed, no experimental evidence that contradicted the predictions of the Standard Model was ever discovered. But in 2001, new experiments began to give tantalizing hints that the Standard Model might not be exactly correct. The experiments were difficult to perform and hard to analyze, and no definite conclusions could be drawn. However, at that time, new and more powerful particle accelerators were under construction and there was hope that results conclusively contradicting the Standard Model might be obtained during the first decade of the twenty-first century.

Yes, particle physicists were hoping that they would discover particle interactions that would demonstrate that their very best theories were not entirely adequate. This is one of the ways in which progress is made in physics. If the predictions of a theory were never contradicted, there would be nothing left to do. On the other hand, when it is discovered that a theory doesn't quite work, new discoveries are sure to lie ahead. Science, after all, is not a body of rigidly held dogma. The "golden ages" of science are the ones during which scientists are confronted with phenomena they don't quite understand. This compels them to invent ways to compel nature to reveal more of its secrets so that a greater degree of understanding can be achieved.

Infinity Problems in Quantum Mechanics

Quantum mechanics is one of the most successful scientific theories ever discovered, and it is the theory on which virtually all of modern physics was based. Yet there is something about the theory that is unsettling. Shortly after quantum mechanics was discovered, physicists began to try to develop a *quantum field theory* that would describe the interactions between light (and other kinds of electromagnetic radiation) and matter. It wasn't so difficult to develop such a theory. However, there was a big problem. Quantum field theory predicted that certain quantities, such as the mass and charge of the electron, were infinite. Something was clearly wrong. These quantities are actually very small.

A solution to the problem was found in 1948 by three physicists working independently, the American physicists Julian Schwinger and Richard Feynman and the Japanese physicist Shin'ichiro Tomonaga. Initially the methods used by the three physicists seemed to be very different from one another. However, they were soon shown to be mathematically equivalent. Schwinger, Feynman, and Tomonaga had found a way to "subtract out" the infinities, making the theory work.

The method of getting rid of the infinities is called *renormalization*, and the theory that results when the infinities are gotten rid of is called *quantum electrodynamics* or QED. QED has proved to be an extraordinarily successful theory. In fact, it is part of the Standard Model. Yet there is one disquieting thing about it—renormalization may not be a mathematically legitimate procedure. Although Feynman was one of the co-inventors of renormalization, he always considered it to be suspect. According to Feynman, "Having to resort to such hocus-pocus has prevented us from proving that the theory of quantum electrodynamics is mathematically self-consistent . . . I suspect that renormalization is not mathematically legitimate." When Feynman was in a jovial mood, he sometimes expressed this idea more succinctly, saying that he had received the Nobel Prize for "sweeping some infinities under the rug."

The Standard Model depends on two theories modeled on QED. One is called *quantum chromodynamics*, or *QCD*. QCD describes the strong force and the interactions between quarks. The other theory, known as the *electroweak* theory, is a generalization of QED. It is a unified theory that describes the interactions between particles that are subject to the weak and electromagnetic forces. Both QCD and the electroweak theory yield predictions of infinite quantities, and in both cases the infinities are removed by the process of renormalization.

Both theories give can be used to do calculations that produce results that are in excellent agreement with experiment. Without them it would never have been possible to create the Standard Model. But as successful as the theories are, something is disquieting about them. They both depend upon a mathematically suspect procedure.

It appears that the basic theories on which the Standard Model depends may not only be approximate descriptions of nature, they may also be theories that are inherently illogical. It may be that they work as well as they do only because three physicists independently found a way to "sweep some infinities under the rug." These are obviously not the kind of theories Einstein was thinking of when he asked whether "God had any choice when he created the universe." Einstein's own theories always had a compelling inner logic. QCD and the electroweak theory possess nothing of the sort, however well they may work in practice.

Superstring Theory to the Rescue?

QCD and the electroweak theory explain three of the four forces of nature. This seems to suggest that it might be possible to create a quantum theory of gravity along the same lines. The theory might produce infinities similar to those found in QED, QCD, and the electroweak theory. But surely it would be possible to remove them by a renormalization process?

The answer to this question is: No, it isn't possible. Attempts to create such a theory were made long ago. Physicists soon gave up on them. If one attempts to combine quantum mechanics and general relativity in any reasonably straightforward way, the resulting theory produces infinities that are worse than the infinities in the theories I have previously described. They cannot be removed by renormalization, and no one knows of any alternative procedure that might eliminate them.

The reason is that gravity is a more complex force than the other three. According to general relativity, such massive bodies as stars, planets, and galaxies are not the only source of gravity. There is energy in a gravitational field, and this energy produces additional gravitational force. One could say that "gravity gravitates." It is this added complication that produces the intractable infinities.

There is hope that superstring theory might eventually solve all these problems. Not only do superstring theories automatically include gravity, they may also eliminate the need to resort to renormalization. Even though superstring theorists have discovered only approximate equations—though they do not know what the mysterious M theory is—they may eventually produce a theory that not only explains all four of the forces, including gravity, but which also eliminates the need to resort to a questionable mathematical procedure.

This again brings up the question of whether superstring theory, unlike all other theories in physics, would be a final theory, one that gave an exact description of nature, not an approximate one. This is a fascinating question. However, it might take decades before anyone knows the answer to it. It may be that physicists will never discover a theory that describes nature exactly. They may forever find themselves discovering theories that are superior to their predecessors, and never reach the end of their quest. Nature may indeed be like an onion with an infinite number of layers that must be laboriously peeled off one by one.

On the other hand, superstring theory (or M theory) may provide some final answers. It may allow physicists to gain a perfect understanding of the fundamental processes that takes place in nature. If this happens, physics will not come to an end. Such a theory would be a theory of fundamental particles and forces. It would not help scientists gain a better understanding of such phenomena as the melting of iron at high temperatures, or air turbulence, or what happens when an electron encounters a helium atom, or any of thousands of other things. However, if a final theory is eventually discovered, then scientists will no longer be dealing only with approximations. They will have discovered the final truth about the particles and forces that make up our universe.

Naturally, no one knows whether or not this will ever happen. In no science are there found anything but approximate descriptions of nature. Yes, physicists know that neutrons and protons are each made up of three quarks. But they describe the interactions between the quarks only in approximate ways. Yes, molecular biologists understand the structure of DNA. But DNA is a template for the manufacture of proteins within biological cells. Biologists are far from an understanding of this process. They don't know how many proteins the DNA within a typical cell makes. They don't know what functions most of the proteins perform, or how they interact with one another. Knowledge of the processes that go on within cells is approximate and incomplete.

Yes, scientists know that the universe began with a big bang. But their theories say nothing about what was happening when the universe was less than 10^{-43} seconds old, and the theories that are used to describe what happened after that time use approximations as much as any theory in science.

As time passes, science comes to have a better and better grasp of how our universe works. However, there is no end in sight. And there may never be an end in sight unless the superstring theorists really do discover some final theory.

Afterword: The Big Questions

Science seems to have a lot to say about some of the age-old questions that were originally posed by philosophers. In particular, investigations carried out in the fields of physics, cognitive science, and evolutionary biology have shed a great deal of light on questions about the nature and origin of the universe, on human nature, and on the workings of the human mind. But science has not come up with any final answers in any of these cases. However much is learned, puzzles always remain.

Scientists have not penetrated all the mysteries of time. In particular, questions concerning time asymmetry remain puzzling. Nothing could be more obvious than that time has a direction; past and future are two very different things. But this has not really been reconciled with the fact that the fundamental laws of physics are time symmetric, that they would have the same form if the direction of time were reversed. Certain physical phenomena allow physicists to define "arrows" of time. For example, the entropy of the universe is increasing; it was lower in the past and it will be higher in the future. Electromagnetic radiation travels toward the future, not the past. The universe is expanding. Thus there will have been greater expansion in the future than there was in the past. But physicists don't know what these arrows of time have to do with one another, or whether there are any connections between them. For example, no one is entirely sure whether or not time would run backwards in a contracting universe. No one is certain whether, under some circumstances, electromagnetic radiation *could* propagate into the past. Scientists don't know why two particles, the kaon and the B meson should be sensitive to the direction of time, while other particles are not. Finally, they don't know how any of the arrows of time defined by physicists are related to our subjective sense of time.

Subjective time may be the most mysterious phenomena of all. We all have the feeling of journeying toward the future. Time is something that "passes." But whatever subjective time is, it is not the time of physics. Physics treats time as a dimension. It is possible to speak of the time interval between two events, just as it is possible to speak of the distance between two objects. But physics knows nothing of the "flow" of time; it is something that cannot be measured. At best one could say that time passes at the rate of one second per second, but that is so tautological it is meaningless. Like St. Augustine, we know what time is until someone asks us. Then we discover we don't know what it is at all.

Nor have the ancient questions about determinism and free will been answered by science. We have the subjective feeling that we are free to act as we choose, and this subjective feeling is probably an accurate one. However, human minds, like everything else in the world, are made of matter. And objects that are too large to be affected by quantum randomness behave in basically deterministic ways. The human mind is made of neurons, which are much too big to be sensitive to any underlying quantum indeterminacy. And if quantum events did affect the brain somehow, it doesn't look as though this would allow us any freedom. At most, random quantum events could only create a kind of "white noise."

The many worlds interpretation of quantum mechanics opens up the possibility that there are an infinite number of coexisting universes. If this interpretation is correct, we may make *all* possible choices in one universe or another. However, the many worlds theory is only one of several possible interpretations of quantum mechanics, and it is not certain whether or not scientists will ever be able to perform experiments that could determine whether or not it is correct. It has been suggested that quantum computers might be able to detect the existence of these alternate universes, but until the experiments are performed, no one will know if this hypothesis is a reasonable one. Since quantum computers are not likely to be created for a long time to come, it doesn't appear that answers are going to be found anytime soon.

Science has learned a lot about the basically deterministic behavior of macroscopic objects and about random events in the quantum world as well. But the knowledge gained has not really thrown any light on questions about determinism in human life. Questions about free versus determinism are still metaphysical ones, and best left to those with a taste for pondering such things. Nowadays it might seem more reasonable to ask, "How is it that we are free?" than it would be to maintain that our lives might be determined from the moment of our births. However, this is still a question best left to the philosophers.

Since the time of Plato, philosophers have been asking questions about the reality of the material world. Indeed there have been some, like Bishop Berkeley, who maintained it wasn't real. During the twentieth century, this particular kind of metaphysical questioning became unfashionable, at least among philosophers. Thus it was the physicists who began to explore such questions instead. According to the first interpretation of quantum mechanics to be developed, the Copenhagen interpretation, subatomic particles do not possess such objectively real properties as position and momentum (or velocity) until they are observed. The formulation of this interpretation set off a debate that is still going on today.

The discovery of Bell's theorem and experiments done by the French physicist Alain Aspect and others put an end to some of the arguments about quantum mechanics by demonstrating that subatomic particles do not have the property of local reality. This means that either they do not possess objectively real properties or that causal influences acting between them are propagated at velocities greater than that of light. Most physicists find the second possibility unappealing. According to Einstein's special theory of relativity, signals that traveled faster than light could also travel from the future into the past. Few of them accept this as a real possibility (although there is one interpretation of quantum mechanics that is based on the idea that this is possible). The majority of physicists conclude that there is a sense in which such particles as electrons and photons are not objectively real.

Experiments performed near the end of the twentieth century showed that if this is indeed the case, then there are also circumstances under which macroscopic objects do not possess full objective reality, either. The experiments showed that superconducting rings large enough to be seen with the naked eye could exhibit quantum behavior. Electrical currents were seen to move around the rings in both directions at the same time. This did not mean that half of the electrons in the rings moved in one direction and half in the other; they were moving in both directions at the same time. The old idea that quantum effects could be seen only in the subatomic world was disproved. In a sense it was as though someone had seen Schrödinger's cat in a half-alive, half-dead state. Recall that Schrödinger's thought experiment was supposed to be paradoxical because it was thought that a macroscopic object like a cat could not exist in a mixture of two quantum states. That idea was shattered when the superconductor experiments were performed. Thus there may be a sense in which even the macroscopic world does not possess full objective reality.

There is one interpretation of quantum mechanics that restores this reality—at a price. That is the many worlds interpretation of quantum mechanics. According to this interpretation, every quantum alternative is realized in one parallel universe or another. Schrödinger's cat is alive in some universe and dead in others, and if electrical currents are seen flowing both ways in a superconductor, this is only an indication that two universes have temporarily come together. The many worlds interpretation is unappealing to some physicists, such as John Archibald Wheeler, who object that it carries "too much metaphysical baggage." But others, such as Stephen Hawking, endorse this interpretation enthusiastically.

Since experiments cannot distinguish between one interpretation of quantum mechanics and another, the debate among physicists is fundamentally philosophical in nature. Questions about the reality of the external world were once argued about by philosophers. Nowadays it is the physicists who have taken up such questions. But they are no nearer a final solution than the philosophers ever were.

Modern cosmology has discovered plausible answers to the question, "Why is there anything at all?" Since gravitational energy is negative, and appears to balance out the matter in the universe, the universe could plausibly have contained no matter at all originally. Thus it is possible that the universe began as a quantum fluctuation of some kind. It is also conceivable that it may have budded off from some previously existing universe.

However, all of the various theories about the origin of the universe have remained speculative. There are as yet no theories in physics that can describe what happened at the moment of creation. When the universe was a tiny fraction of a second old, it was very hot and dense. Conditions were so extreme that only a theory of quantum gravity could describe the processes taking place. But no theory of quantum gravity exists. Quantum mechanics and Einstein's theory of gravitation—his general theory of relativity—are incompatible. There is some hope that superstring theory, which can account for all four of the forces of nature, including gravity, might eventually allow cosmologists to understand what was happening when the universe was less than 10^{-43} seconds old. The physicists and mathematicians who work with superstring theory have so far only been able to find approximate solutions to approximate equations. They know that the five different superstring theories are all related to M theory. But no one yet knows what M theory is, or on what principles it might be based. Scientists in the field of superstring cosmology have produced some intriguing ideas about the prehistory and origin of the universe, but these hypotheses are even more speculative than the scenarios of the origin of the universe that do not depend upon superstring theory. There is simply too much that the superstring theorists do not yet understand.

Einstein once asked whether God had any choice when he created the universe. When he posed this question, Einstein was speaking metaphorically. What he was asking was whether the requirement of logical consistency alone dictated the form of the laws of nature or whether they could conceivably have been different.

Einstein's question has not yet been answered. Some physicists hope that superstring theory will eventually advance to the point where it will explain why the four forces of nature have the specific strengths they do, why certain kinds of particles exist, and why they have the properties they have. If the superstring theorists ever do discover such a theory, it is likely they will do so only after decades of theoretical research. During the early twenty-first century no one knew whether physical laws and physical constants had the character they did because they *had* to be that way, or whether the character of physical laws and constants was in part a matter of chance.

Scientists do not know whether a final theory, one that explained the character of the laws of nature, will ever be discovered. It may be that this will eventually happen. Or it may be that theoretical research in physics will always be something like peeling away the layers of an onion. It is conceivable that physicists will continue to gain a better understanding of the laws of nature for many generations to come, but that they will never reach the end of their quest. Some scientists hope that a final theory will eventually be discovered. But others suspect that there is no such thing and that our understanding of nature will always be imperfect.

Cosmologists today frequently ask the question, "What happened before the big bang?" They normally don't phrase the query that way. However, there is a lot of speculation about what, if anything, might have existed before our universe began. In the past cosmologists often evaded the issue by speculating that both space and time were created in the big bang. Now, they frequently construct scenarios in which big bangs take place in widely separated regions of previously empty space, or in which new universes are created in ones that already exist.

In ancient times, the cosmos was conceived of as a thing of limited size. The ancient Greeks, for example, believed that it consisted of Earth and seven planets—the sun and moon were counted as planets—surrounded by a sphere in which the stars were embedded. The heliocentric theory of Copernicus enlarged the universe considerably. If Earth did indeed revolve around the sun, as Copernicus concluded, then it was conceivable that the stars were, in fact, other suns. The cosmos might be vast indeed. Discoveries made in subsequent centuries enlarged the universe even more. Until around the beginning of the twentieth century, astronomers believes that our Milky Way galaxy *was* the universe. Other galaxies were seen by astronomers, but since telescopes were not powerful enough to resolve the individual stars in them, it was thought that they were nothing more than clouds of gas. But all this changed when powerful new telescopes were constructed on Mount Wilson and Mount Palomar in California. Using these telescopes, astronomers discovered that our galaxy was only one of a great number. Some of the objects that their predecessors had seen did turn out to be gas clouds. But many others were galaxies as large as, or larger than, the Milky Way. Today astronomers know that our universe is vast indeed. It is estimated that there are about 100 billion galaxies in the observable universe.

Nowadays, scientists envision a cosmos that is incredibly vast. There is no particular reason to believe that our universe is the only one that exists. If our universe did begin as a quantum fluctuation, or of it budded off from some previously existing universe, other universes might easily have been created in the same way innumerable times. Thus cosmologists commonly speak of the multiverse, a cosmos containing a very large number, possibly an infinity, of universes.

The idea that a multiverse exists is highly plausible, but it is a hypothesis that has no empirical confirmation. It is natural to speculate that many universes exist. This idea depends on no special assumptions. And if one is to maintain that our universe is the only one, then assumptions of one kind or another have to be invented; there would have to be something that would have prevented the creation of other universes. And no one knows what this "something" could be.

The Renaissance philosopher Giordano Bruno believed that there were an infinite number of worlds and was burned at the stake for this and other heresies. Today the cosmos of Bruno seems a limited one. There are plausible reasons for believing that there is far more to the cosmos than he thought. There are almost too many ways in which universes could be created; a moderately large number of different scenarios have been suggested, and it doesn't seem likely these could all be true.

The existence of other universes has not been experimentally confirmed. If they exist, it is not likely there is any way we could observe them. And, of course, scientists are hampered that there is as yet no theory that can describe what happened at the moment of creation. The idea that a multiverse exists is plausible, but only when a theory is found that can describe the events that took place when the universe was less than 10^{43} seconds old will scientists be able to put the hypothesis of the multiverse on a sound scientific footing.

The universe in which we live seems an improbable one. It is almost as though physical laws and constants have been fine-tuned to make the emergence of life inevitable. Small changes in the strengths of any of the four fundamental forces, or equally small changes in any of a number of physical constants would produce a universe that was inhospitable to life. To the modern proponents of the argument from design, this is evidence for the existence of an intelligent designer, often identified with the Judeo-Christian creator. But many scientists dispute the validity of this argument, pointing out that it is likely many other universes exist. Furthermore, they say, if the laws of nature can vary from one universe to another, then most of them would be devoid of life. The reason our universe has the character it has is that if it did not there would be no one here to see it.

The question of the validity of the argument from design is no more settled than it was two hundred years ago. Scientists do not yet know if physical laws could be other than what they are, or whether physical constants could vary from one universe to another. This is one of the questions that may eventually be answered by superstring theory. However, no one expects that it will be any time soon.

It is not likely that the arguments about intelligent design will end. According to such scientists as Stephen Jay Gould, religion and science are "non-overlapping magisteria" that deal with different areas of human experience. But this is disputed both by atheists such as Richard Dawkins and by religious philosophers such as Richard Swinburne. Dawkins claims that religion does make claims that could in principle be empirically judged to be true or false. He gives the virgin birth as an example. Swinburne, on the other hand, maintains that evidence of design is there for anyone to see, and cites our ability to understand the universe as evidence of intelligent design.

Human beings generally look for things that can give meaning to their lives. In past ages, some of them found this in science, even though they weren't scientists themselves. During the Victorian age, belief in the value of progress was so fervent that it was almost a religion. Science was valued because it was thought scientific discoveries would help to better the human condition. Indeed Science has done this. However, the world picture science presents to us today seems anything but meaningful. We live in a universe that came into existence by chance, possibly as the result of a quantum fluctuation. The universe was initially hostile to life. There were initially no elements heavier than hydrogen and helium, and thus no planets on which life could begin to evolve.* Billions of years from now, when it begins to grow cold and dark, our universe will again become inhospitable to living things, and life will eventually die out.

Evolutionary biology presents us with a picture that seems no more meaningful. The design we see in the natural world was created by natural selection, not by a beneficent creator. Human beings are not the final product of evolution, in any sense of the term. Evolution followed the paths it did partly because of such chance events as mass extinctions, and the evolution of intelligence was most likely not inevitable. It may very well be that intelligence never arises in most of the places in the universe where life evolves. Millions of different species have evolved on our planet, but intelligence arose in only one of them. As the German-American biologist Ernst Mayr points out, the mammals consist of some twenty-odd orders, but the evolution of intelligence occurred in only one of them. There are over a hundred species of primate, but again, intelligence arose only once. Mammals are only one of a very great number of evolutionary lines that have evolved on Earth. The evolution of the kind of intelligence that would be capable of developing a civilization is likely to have been no more inevitable than the evolution of the trunk of the elephant. Although life may exist in many places in the universe, intelligence may exist in very few of them. Indeed, it is possible that ours is the only technological civilization in the universe. I, for one, hope it is not. However, it must be considered a distinct possibility.

---

*Some light elements like lithium were present in minute quantities. But the existence of these elements had no significant effect on the subsequent evolution of the universe.

We exist by chance in a universe that was created by chance. Science that shows us this can hardly be a source of meaning in life, except for scientists who find fulfillment advancing human knowledge. We devoutly wish for lives that are meaningful, but it appears it is necessary to find meaning elsewhere. There is little or none in the picture that science draws of our universe.

Human beings have always been interested in the character of human nature. Most of us are interested in knowing why we behave the ways we do. And, of course, assumptions about human nature have to be made if one is to develop a system of ethics. For most of the twentieth century, the idea propounded by cultural anthropologists that culture was overwhelmingly important was dominant. But this idea is now being challenged by scientists working in the relatively new discipline of evolutionary psychology. The evolutionary psychologists believe that there are universal human behavioral traits, and these traits evolved because they were adaptive in the environments in which our ancestors lived. In other words, they dispute the notion, put forward by cultural anthropologists, that there is no such thing as basic human nature.

Evolutionary psychology has been harshly criticized by such scientists as Stephen Jay Gould. However, it has become reasonably clear that at least some of the theories advanced by evolutionary psychologists are valid. For example, it has been shown that mate preferences contain elements that do not vary from one society to another. Convincing arguments have been put forward that pregnancy sickness is not a side effect of hormonal changes, but that it is an unconscious behavior that evolved because it was adaptive in our ancestral environment. It is possible that there are not as many genetic influences on our behavior as some of the evolutionary psychologists believe, but it has become obvious that such influences do exist. As a result, scientific conceptions of human nature are changing. It remains to be seen how much they will change. It is clear, however, that scientific investigations are giving us more insights into ourselves.

"What is mind?" could almost be called an archetypal philosophical question. Philosophers have attempted to answer it in one age after another. In our time, it has become a scientific question, and cognitive scientists are beginning to achieve some significant results. Cognitive scientists make use of the functionalist theory of the mind because it can help them to make new scientific discoveries. According to functionalism, the mind is what the brain does. Mind, in other words, is nothing more than a succession of brain states.

Cognitive scientists frequently make use of computer models. They believe that if a computer can be made to accomplish something that would be called "intelligent" if a human being had done it, they have gained an understanding of some of the things that go on in human minds. The widespread use of computer models has led in turn to a great deal of speculation about the creation of computer intelligence. If the functionalist theory is correct, then it should be possible for intelligence to exist in silicon chips as well as in biological brains.

This claim has led to a lot of philosophical argument about the possibility of computer intelligence. Some philosophers claim that there are fundamental reasons why it is impossible, while computer scientists more often argue the opposite. The question of whether or not computers can be intelligent is an empirical one. But it is one that cannot yet be answered. It is likely that years or decades of research will be required before it can be. Until it can be answered, scientists will not be able to fully answer the related question, "What is mind?"

There also remain unanswered questions about the nature of scientific knowledge itself. The most highly developed of the physical sciences is physics, but physics is a science of approximations. No physical "law" or theory gives an exact description of reality. For example, Newton's law of gravitation has turned out to be a special case of Einstein's general theory of relativity, one that is valid under certain conditions. General relativity is a special case of—and thus an approximation of—some theory not yet discovered. Physicists hope that we will eventually develop a quantum theory of gravity, which will tell them how the force of gravity behaves when quantum effects become important.

This raises the question of whether scientific knowledge will always remain approximate, or whether scientists will eventually discover a theory that gives an exact description of the properties of the fundamental particles and forces that exist in our universe. Some scientists hope that superstring theory (or M theory) will provide such a description. Others suspect that a final theory will never be discovered, that better and better descriptions of nature will be found, and that the search will never come to an end. In other words, no one knows whether or not a complete scientific understanding of our universe is possible.

Asking Questions

It would seem that science has not yet fully answered *any* of the questions I have posed. Physicists do not yet have a complete understanding of the nature of time. They don't know precisely how the universe came into existence, or whether other universes might exist. They do not yet have a complete understanding of the most successful modern theory of physics, quantum mechanics. They are still arguing about how quantum mechanics should be interpreted, and what the implications are for the nature of subatomic reality. They are not sure why there are three dimensions of space. They do not even know whether it will eventually be possible to discover a final theory, one that gives a complete description of the particles and forces that are observed in our universe.

Cognitive scientists have made great progress toward attaining an understanding of the human mind. But there are many unanswered questions here, too, including the one of whether a mind could exist in something other than a biological brain. Scientists don't fully understand human nature. Anthropologists have shown that we are molded in part by our cultures, and evolutionary psychologists have demonstrated that human nature also has a genetic component. The latter are only beginning to work out the details, and some of their ideas have been harshly criticized.

Science has shown considerable light on these and many other fundamental questions: if there is much yet to be discovered, that is no tragedy. Science, like philosophy, is a method of questioning. Philosophers initially posed many of the questions that scientists are now attempting to answer. Scientists try to answer them by putting questions to nature, by making accurate observations and by performing carefully designed experiments.

Perhaps there are no questions that are wholly philosophical in nature. If a question can be meaningfully asked, then it should be possible to make attempts to answer it empirically. Naturally some questions are deeper than others, and thus more difficult to answer. But the difficult questions are frequently fundamental ones. And I suspect that some of the answers will be very surprising. They are often the "big questions" that human beings have been trying to answer, in one way or another, for more than two thousand years.

Glossary

absolute zero. The lowest possible temperature, at which all molecular motion ceases. It is equal to –273 degrees Celsius.

advanced radiation. Radiation that travels from the future into the past. Advanced radiation is allowed by the laws of electromagnetism, but it is not observed.

antiparticle. Every known particle has an antiparticle. The antiparticle of the electron is the positron, which has properties identical to those of the electron, but which is positively charged. When a particle and its antiparticle meet, they mutually annihilate each other in a burst of energy.

arrows of time. An arrow of time is any physical process that can be used to define the direction of time. Since entropy increases, we can say that the direction of increasing entropy is an "arrow" pointing toward the future.

baby universe. In some theories new universes are supposed to "bud off" from previously existing universes. The newly formed "buds" are called baby universes.

big bang. The universe is thought to have begun in a very hot, very highly compressed, and rapidly expanding state. This is called the big bang.

big crunch. If, over a period of billions of years, gravity were able to halt the expansion of the universe, a phase of contraction would follow, and the universe would eventually collapse in a big crunch. A big crunch would be analogous to the big bang except that it would be a collapse rather than an expansion.

black hole. A highly compressed remnant of a dead, massive star. Gravity in a black hole is so strong that nothing, not even light, can escape it. Hence the name "black hole."

boson. A particle that transmits forces. For example, electromagnetic force is transmitted by the photon, which is also a particle of light.

brown dwarf. Bodies that are not quite massive enough to become stars. Pressures and temperatures in their interiors do not become high enough to cause the nuclear reactions that power stars to begin.

celestial mechanics. A field of physics devoted to the study of the motions of astronomical bodies.

chaotic inflation. A version of the inflationary universe theory postulates that big bangs like the one that produced the universe take place at many different times and places in a kind of super-universe that is immeasurably vast. This process is known as chaotic inflation.

dark energy. A hypothetical energy that fills all space. According to Einstein's equation $E = mc^2$ this energy would have the same gravitational effects as an equivalent quantity of matter.

dark matter. Matter that is present in the universe that astronomers cannot observe directly. It is known to exist because its gravitational effects can be seen.

deuterium. Sometimes called "heavy hydrogen." The nucleus of a deuterium atom consists of a proton and a neutron, while the nucleus of an ordinary hydrogen atom contains only a proton.

electromagnetic radiation. Light, radio waves, X rays, gamma rays, and infrared and ultraviolet radiation are all made up of oscillating electric and magnetic fields. They are referred to collectively as electromagnetic radiation.

electroweak theory. A unified theory that explains both the electromagnetic and the weak nuclear force.

entropy. See *Second Law of Thermodynamics*.

eukaryotic. A biological cell that contains a nucleus. Plants and animals are made of eukaryotic cells. These cells are considerably more complex than those of bacteria, which have no nucleus, and which are called prokaryotic.

expanding universe. The galaxies and clusters of galaxies that make up the universe are moving farther apart from one another. This is called the expansion of the universe.

fermion. A particle of matter. Examples of fermions are electrons and the quarks of which protons and neutrons are composed.

functionalism. A theory of the mind that avoids the problems associated with dualism of "mind" and "body" by assuming that "mind" is what the brain does. According to this theory, then, "mind" is nothing more than a succession of different brain states.

gluon. Force particles that bind quarks together in protons and neutrons. Gluons are also responsible for the forces that hold atomic nuclei together.

Heisenberg uncertainty principle. One of the fundamental principles of quantum mechanics, it states that the position and momentum (or velocity) of a particle cannot be simultaneously determined.

inflationary universe theory. The theory that when the universe was a tiny fraction of a second old there was a brief period in which its rate of expansion was far greater than it was either before or after this time. The theory is accepted by most cosmologists.

many worlds interpretation. In quantum mechanics, the many worlds interpretation postulates that there exist a very large, probably infinite number of coexisting universes. According to this interpretation, some of these universes are very different from ours, while others differ only in insignificant ways. The idea resembles the "parallel worlds" concept used by some science-fiction writers.

M theory. It has been shown that the five known superstring theories are all related to a more comprehensive M theory, and are therefore related to one another. Physicists have not been able to deduce what this M theory is.

multiverse. A term that describes the ensemble of different universes in the many worlds interpretation of quantum mechanics.

nucleotides. The molecules in DNA that carry the genetic code.

photon. Quantum mechanics tells us that light has the properties of both waves and particles. Particles of light are called photons.

Planck era. Refers to times when the universe was less than 10^{-43} seconds old. There exists no theory capable of describing the process that took place at these early times.

positron. The positron is the antiparticle of the electron (see *antiparticle*).

QCD. Quantum chromodynamics, the theory that describes the interactions between quarks.

QED. Quantum electrodynamics, the theory that describes the interaction of light and other kinds of electromagnetic radiation with matter.

quantum cosmology. A new field of physics in which quantum mechanics is used to attempt to understand the origin and early history of the universe.

quantum fluctuations. Pairs of particles and antiparticles constantly pop into existence everywhere, even in "empty" space, and then mutually annihilate each other a tiny fraction of a second after they are created. The creation of such a particle-antiparticle pair is called a quantum fluctuation. The "virtual particles" created in this manner cannot be directly observed, but they do produce measurable effects, which confirms their existence.

quantum gravity. A theory that combined quantum mechanics and Einstein's general theory of relativity (his theory of gravitation) would be a theory of quantum gravity. But quantum mechanics and general relativity are mutually incompatible, and no such theory has ever been found.

quantum jump. When an electron in an atom goes from one energy state to another, emitting a photon of light in the process, it is said to undergo a quantum jump.

quark. The components of such particles as neutrons and protons. Each particle is made up of three quarks.

relativity, special and general theories of. Einstein's special theory of relativity deals with objects that move at velocities close to that of light. His general theory of relativity is a theory of gravity.

renormalization. A mathematical procedure that is used to remove the infinities that arise in such theories as QED and QCD. Renormalization may not be a mathematically consistent method, but it produces results that are confirmed by experiment to extremely high degrees of accuracy.

Second Law of Thermodynamics. The entropy (disorder) of an isolated system must always increase. The words "isolated system" are important here. For example, the Second Law does not apply to Earth, which is not isolated, since it is constantly receiving energy from the sun.

singularity. According to Einstein's general theory of relativity, the matter inside a black hole is compressed into a mathematical point of infinite density called a singularity. It is likely that the existence of such an extreme state is prevented by quantum effects. No one knows what these quantum effects would be, since relativity generally cannot be combined with quantum mechanics.

spacetime. The three dimensions of space and the dimension of time are collectively known as spacetime. Spacetime is four-dimensional. (Note that this is not the same as four dimensions of space.)

spin. A property of subatomic particles that bears some resemblance to the spin of macroscopic objects. However, since it is a quantum effect, there are important differences.

Standard Model. A set of theories developed in the 1970s to explain the properties of the fundamental particles and forces. At the end of the twentieth century, no experiment had ever contradicted the Standard Model.

strong force. The force that bind quarks together in protons and neutrons, and which binds protons and neutrons together in atomic nuclei.

superconductor. In some substances there is no electrical resistance at low temperatures, hence any electric currents that are set up in them are maintained without any additional expenditure of energy. Such substances are called superconductors.

supernova. When the nuclear fuel of a very large star is exhausted, it undergoes a supernova explosion. Supernovas are quite spectacular. For a short period they can shine as brightly as an entire galaxy.

supersymmetry. An abstract mathematical symmetry that would describe the relationship between fermions (matter particles, see *fermion*) and bosons (particles of force, see *boson*). It has not yet been demonstrated that supersymmetry exists in nature. However, during the early years of the twenty-first century, new, more powerful particle accelerators were being constructed, and physicists hoped that experiments performed with them would demonstrate that supersymmetry existed.

three-body problem. The problem of determining the motion of three massive bodies that attract one another gravitationally cannot be solved exactly. Hence approximate methods must be used if this motion is to be calculated.

time reversible. All of the fundamental laws of physics are time reversible. They do not distinguish between the two possible directions of time, and would still be valid if time "went backwards" (if past and future were reversed).

Turing test. A test proposed by the British mathematician Alan Turing. According to Turing, if a person who communicated with a computer by typing at a computer console could not tell whether he was talking to a human being or a machine, then the computer could be said to be "intelligent." The validity of the Turing test is widely, but not universally, accepted.

virtual particles. See *quantum fluctuation*.

weak force. The force responsible for certain types of nuclear decay. The name is a reference to the fact that it is much weaker than the strong force, which holds nuclei together. However, the weak force is quite strong compared to the force of gravity.

white dwarf. When a small- or medium-size star exhausts its nuclear fuel, it shrinks into a white dwarf, which continues to glow only because of its residual heat.

wormhole. A structure that would connect two different regions of spacetime. Although wormholes have never been observed, they are theoretically possible according to Einstein's general theory of relativity.

For Further Reading

All of the following books should be accessible to the general reader:

Brockman, John. *The Third Culture*. 1995. New York: Simon & Schuster. This book is based on interviews with scientists Stephen Jay Gould, Richard Dawkins, Roger Penrose, Alan Guth, Murray Gell-Mann, and others.

Buss, David M. *The Evolution of Desire*. 1994. New York: Basic Books. An evolutionary psychologist looks at human mating patterns.

Campbell, Jeremy. *The Improbable Machine*. 1989. New York: Simon & Schuster. Not the most up-to-date book on artificial intelligence, but highly readable.

Churchland, Paul M. *Matter and Consciousness*, revised ed. 1988. Cambridge, Mass.: MIT Press. Churchland is a philosopher with a good understanding of the cognitive sciences.

Conveney, Peter, and Roger Highfield. *The Arrow of Time*. 1990. New York: Ballantine Books.

Crevier, Daniel. *AI*. 1993. New York: Basic Books. A history of artificial intelligence.

Davies, Paul. *About Time*. 1995. New York: Simon & Schuster.

Davies, P. C. W., and J. R. Brown, eds. *The Ghost in the Atom*. 1986. Cambridge, Eng.: Cambridge University Press. The editors interview proponents of a number of different interpretations of quantum mechanics. P. C. W. Davies is the same individual as Paul Davies.

Dawkins, Richard. *The Blind Watchmaker*. 1986. New York: W. W. Norton & Co. The author of *The Selfish Gene* explains how natural selection creates design in nature.

Dennett, Daniel C. *Darwin's Dangerous Idea*. 1995. New York: Simon & Schuster. One of the theses of this book is that many people find Darwin's theory of evolution to be threatening because they fear it eliminates any sense of purpose and meaning in life. Dennett is a well-known philosopher.

Deutsch, David. *The Fabric of Reality*. 1997. New York: Penguin Press. A look at recent thinking about parallel universes.

Ferris, Timothy. *The Whole Shebang*. 1997. New York: Simon & Schuster. A readable account of modern cosmology by a veteran science writer.

Feynman, Richard. *QED*. 1985. Princeton, N.J.: Princeton University Press. A readable account of quantum electrodynamics.

Flood, Raymond, and Michael Lockwood, eds. *The Nature of Time*. 1986. Oxford, Eng.: Basil Blackwell.

Franklin, Stan. *Artificial Minds*. 1997. Cambridge, Mass.: MIT Press.

Glynn, Ian. *An Anatomy of Thought*. 1999. New York: Oxford University Press. A book about the brain and the mind.

Gould, Stephen Jay. *Full House*. 1996. New York: Harmony Books. Gould argues that there is no "progress" in evolution.

———. *Rocks of Ages*. 1999. New York: Library of Contemporary Thought. Gould argues that science and religion are "non-overlapping magisteria."

Greene, Brian. *The Elegant Universe*. 1999. New York: W. W. Norton & Co. The best and most up-to-date book on superstring theory.

Gribbin, John. *In the Beginning*. 1993. Boston: Little, Brown & Co. A British physicist describes some radical new ideas about the universe.

———. *Schrödinger's Kittens and the Search for Reality*. 1995. Boston: Little, Brown & Co.

———. *The Search for Superstrings, Symmetry, and the Theory of Everything*. 1998. Boston: Little, Brown & Co.

Guth, Alan H. *The Inflationary Universe*. 1997. Reading, Mass.: Addison-Wesley. The originator of the inflationary universe theory discusses his theoretical discoveries and those of other cosmologists.

Hogan, James P. *Mind Matters*. 1997. New York: Ballantine Publishing. An account of research in the field of cognitive science.

James, William. *The Principles of Psychology*, Vol. 2. 1950. New York: Dover. In some interesting passages James analyzes the reasons for human beliefs.

———. *Some Problems of Philosophy*. 1996. Lincoln, Neb.: University of Nebraska Press.

Kane, Gordon. *Supersymmetry*. 2000. Cambridge, Mass.: Perseus Publishing. A discussion of supersymmetry, one of the fundamental ideas on which superstring theory is based.

Kosslyn, Stephen M., and Oliver Koenig. *Wet Mind*. 1992. New York: The Free Press. A comprehensive look at cognitive neuroscience.

Maynard Smith, John, and Eörs Szathmáry. *The Origins of Life*. 1999. Oxford, Eng.: Oxford University Press.

Mayr, Ernst. *One Long Argument*. 1991. Cambridge, Mass.: Harvard University Press. An eminent biologist describes the genesis of modern evolutionary thought.

Morris, Richard. *Achilles in the Quantum Universe*. 1997. New York: Henry Holt & Co.

———. *Cosmic Questions*. 1993. New York: John Wiley & Sons.

———. *Time's Arrows*. 1984. New York: Simon & Schuster.

————. *The Universe, the Eleventh Dimension and Everything.* 1999. New York: Four Walls Eight Windows.

Penrose, Roger. *The Emperor's New Mind.* 1989. Oxford, Eng.: Oxford University Press. Penrose explains his own, somewhat idiosyncratic, theory of the mind.

Pinker, Steven. *How the Mind Works.* 1997. New York: W. W. Norton & Co. This may be the best book on evolutionary psychology.

Polkinghorne, John. *Belief in God in an Age of Science.* 1998. New Haven, Conn.: Yale University Press. A British theoretical physicist turned Anglican priest argues that there is no conflict between science and theology.

Rees, Martin. *Before the Beginning.* 1997. Cambridge, Mass.: Perseus Books.

————. *Just Six Numbers.* 2000. New York: Basic Books. This and the preceding book are readable accounts of current thinking in cosmology by a British astrophysicist.

Searle, John R. *The Rediscovery of the Mind.* 1992. Cambridge, Mass.: MIT Press.

Silk, Joseph. *A Short History of the Universe.* 1997. New York: Scientific American Library.

Swinburne, Richard. *Is There a God?* 1996. Oxford, Eng.: Oxford University Press. A British philosopher argues for acceptance of the argument from design.

Index

Bohm, David, hidden variable theories
and, 60–61, 63
Bohr, Niels, 54–59, 62, 68–69
see also "Copenhagen interpretation"
Boltzmann, Ludwig, cycles of time and,
36–37
Boomerang telescope, 81
Bose, Satyendra Nath, 102
bosons, 102
brain:
cheater detection mechanism, 189–91
consciousness and, 214–16
evolution of, 216–17
injury to, 205–6, 215
mental modules, 186–87
mind and, 200–202, 205–7, 217–18, 242
see also mind
neurons and synapses in, 204, 206,
207, 213
parallel processing of, 204, 206–7
redundancy in, 43
brain stem, 215
Brandenberger, Robert, 121–22
branes, 106
Brief History of Time, A (Hawking), 29,
85n, 137
Broca's area, 187
"Bronze Head, A," 11
brown dwarfs, 79–80
evolution of, 163
Bruno, Giordano, 130
Inquisition and, 128–29, 238
Burton, Robert, 129
Buss, David, 180–83, 187, 193
Byzantine Empire, 92

Cambrian explosion, 170
Cambridge Center for Behavioral
Studies, 197
Cambridge University, 140
carbon, 150, 163
beryllium and, 163
carbon dioxide, 166, 167, 170
Carter, Brandon, 154
Casimir, Hendrik, 74
celestial mechanics, 222–23, 225
centrifugal force, 165
CERN (Centre Européen pour la
Récherche Nucléaire), 27

chance, 91–92, 174
creation and, 139, 143, 240–41
chaotic inflation, 116–19, 126, 157
chaotic systems, weather forecasting and,
39–40
chatterbot (chat robot), 194–96
cheaters, 189–91
chemical evolution, 167–68
chimpanzees, 216, 217
Chinese Room, 210–13
chloroplasts, 169
Christianity:
dualism and, 201
infinity of worlds and, 127–29
Cicero, 36, 139
circular time, 35–37
civilization, 182
clairvoyance, 68
clams, 172
cognitive science, 5–8, 199, 241–42
see also mind
collagen, 170
collisions with asteroids:
formation of solar system and, 165
mass extinctions and, 91, 166–67,
171–72
coma, 215
comets, 38, 167
ice on, 166
Coming of Age in Samoa (Mead), 178–80
compacted dimensions, 89–90, 99
completely ridiculous anthropic principle
(CRAP), 155n
computerized scans, 205–206, 215
computers, 194–219, 242
ALICE, 194–96
central processing unit, 203–4, 206,
207
chess, 197, 204, 207, 218
Chinese Room, 210–13
mind and, 203–5
modeling with, 198–200, 207
most human, 196, 197
parallel processing and, 207–8
serial, 206
serial architecture and, 208
that think, 207–13, 218
Turing test and, *see* Turing test
vision systems and, 199, 206–7

About the Author

RICHARD MORRIS, PH.D., is the author of more
than a dozen books explaining the wonders and
intricacies of the scientific world. Among these
are *The Evolutionists, Achilles in the Quantum
Universe, Time's Arrows,* and *The Edges of Science.*
He lives in San Francisco, California.